Parallel Circuit Simulation: A Historical Perspective and Recent Developments

Parallel Circuit Simulation: A Historical Perspective and Recent Developments

Peng Li

Texas A&M University
College Station, Texas 77843
USA
pli@tamu.edu

now
the essence of knowledge

Boston – Delft

Foundations and Trends® in Electronic Design Automation

Published, sold and distributed by:
now Publishers Inc.
PO Box 1024
Hanover, MA 02339
USA
Tel. +1-781-985-4510
www.nowpublishers.com
sales@nowpublishers.com

Outside North America:
now Publishers Inc.
PO Box 179
2600 AD Delft
The Netherlands
Tel. +31-6-51115274

The preferred citation for this publication is P. Li, Parallel Circuit Simulation: A Historical Perspective and Recent Developments, Foundations and Trends® in Electronic Design Automation, vol 5, no 4, pp 211–318, 2011

ISBN: 978-1-60198-514-9

Foundations and Trends® in Electronic Design Automation

Volume 5 Issue 4, 2011

Editorial Scope

Foundations and Trends® in Electronic Design Automation will publish survey and tutorial articles in the following topics:

- System Level Design
- Behavioral Synthesis
- Logic Design
- Verification
- Test
- Physical Design
- Circuit Level Design
- Reconfigurable Systems
- Analog Design

Information for Librarians

Foundations and Trends® in Electronic Design Automation, 2011, Volume 5, 4 issues. ISSN paper version 1551-3939. ISSN online version 1551-3947. Also available as a combined paper and online subscription.

Foundations and Trends® in
Electronic Design Automation
Vol. 5, No. 4 (2011) 211–318

DOI: 10.1561/1000000020

now
the essence of knowledge

Parallel Circuit Simulation: A Historical Perspective and Recent Developments

Peng Li

Department of Electrical and Computer Engineering, Texas A&M University, College Station, Texas 77843, USA, pli@tamu.edu

Abstract

Transistor-level circuit simulation is a fundamental computer-aided design technique that enables the design and verification of an extremely broad range of integrated circuits. With the proliferation of modern parallel processor architectures, leveraging parallel computing becomes a necessity and also an important avenue for facilitating large-scale circuit simulation. This monograph presents an in-depth discussion on parallel transistor-level circuit simulation algorithms and their implementation strategies on a variety of hardware platforms. While providing a rather complete perspective on historical and recent research developments, this monograph highlights key challenges and opportunities in developing efficient parallel simulation paradigms.

Contents

1

Introduction

As one of the most critical forms of pre-silicon simulation and verification, transistor-level circuit simulation (e.g., SPICE) is essential for the design of a very broad range of integrated circuits and systems such as custom digital integrated circuits (ICs), memories, analog, mixed-signal, and radio-frequency (RF) designs [74]. Circuit simulation serves the critical mission of predicting circuit performance and makes it possible to disqualify a failing design for expensive chip fabrication. Equally, if not more importantly, the ability of predicting circuit performance through simulation is at the core of any design process; it makes the implementation of complex integrated circuits technically feasible and economically viable while relaxing any heavy need for prototyping.

As a fundamental design and verification enabler, circuit simulation has been under a long-lasting active development since the dawn of the semiconductor industry [10, 62, 70, 71, 73, 74, 83, 102, 105, 114]. However, circuit simulation does not come without cost. Several decades ago, when computer hardware and CPU time were both precious commodities, large-scale circuit simulation was an expensive necessity [83]. To date, designers have the access to much cheaper and more powerful computing facilities, thanks to the remarkable growth of the semiconductor industry in the past decades. Some form of circuit simulation is almost indispensable for any IC design flow. Electronic circuit simulation techniques have even diffused into other disciplines for purposes such as predicting the behavior of biological systems [67].

1.1 Historical and Recent Technological Drivers for Parallel Circuit Simulation

While a wealth of simulation methods and tools have been crafted to serve many different simulation needs, there is an unsaturated demand for higher accuracy, faster speed, and new analysis capabilities in simulation as the designers strive to deliver increasingly complex IC designs with more functionality and improved performance. Simulation of large IC designs with increasingly complex device models, necessary for modeling modern fabrication processes, remains as a significant challenge. It is by no means surprising that circuit designers may have to spend days, weeks, or even months of CPU time to perform expensive transistor-level circuit simulation. As such, simulation is one of the most critical bottlenecks in the current design flow. Insufficient simulation limits the extent to which pre-silicon verification and design space exploration may be conducted, contributing to long design turnaround time, sub-optimal designs, and even chip failures.

To this end, it is logical not only to develop efficient simulation algorithms but also to leverage parallel computing to enable large-scale circuit simulation. It is also not surprising that *parallel circuit simulation* is not a new concept. There have been earlier attempts to develop parallel simulation capabilities on vector machines, multi-processors, and supercomputers, either custom built or commercially available [12, 13, 18, 51, 98, 112, 118, 125].

On the other hand, the recent industry's shift to multi- and many-core processor technology has literally made every modern-day desktop, server, and laptop a parallel computing system [3, 8, 25, 34, 36, 45, 50, 64, 72, 85, 96, 109, 116]. This shift toward chip multiprocessors (CMPs) reflects the fundamental performance and power tradeoffs in lieu of VLSI technology scaling. In addition, commodity many-core graphics processing units (GPUs) [4, 77, 78] and heterogeneous processors [2, 86] with impressive computing power have also merged. For instance, modern GPUs may integrate hundreds of streaming processors on-chip, deliver a peak performance of hundreds of GFLOPS to several TFLOPS, and have a memory bandwidth exceeding 100 GB/s [4, 78]. User-friendly programming model and interfacing tools have also merged for developing general-purpose applications on GPUs [77].

1.2 New Challenges and Opportunities of Parallel Circuit Simulation

This change of the computing landscape has produced profound implications on how compute-intensive applications shall be developed. It has also sparkled active new research activities on parallel circuit simulation. The motivations behind this renewed interest in parallel circuit simulation are multifold.

As power consumption acts as a show stopper for continuing scaling the clock frequency, the performance of single-threaded applications quickly saturates. Simply put, the software developer's "free ride" of the Moore's Law is coming to an end. Further improvement of software performance may only come from explicit exploration of parallelism [6]. To certain extent, this fact has forced the researchers and practitioners to more seriously look into parallel simulation. More importantly, it has stimulated active and exciting development of modern commercial parallel circuit simulators from all major EDA tool vendors and across the industry (see e.g., [1, 32, 110]), and provided impetus for new research in parallel circuit simulation and other areas of parallel electronic design automation [11].

On the other hand, advancements in CMOS and processor technologies have made parallel computing ubiquitous. Modern multi- or many-core processors offer an amount of compute power that was even hardly available on bulky and expensive mainframes and supercomputers decades ago. Today, affordable terascale parallel compute power is at the disposal of typical circuit designers. As illustrated in Figure 1.1, a diverse spectrum of parallel hardware platforms exists, ranging from CMPs, heterogeneous processors, hardware accelerators (GPUs and FPGAs), and SMPs to computer clusters and supercomputers. Each

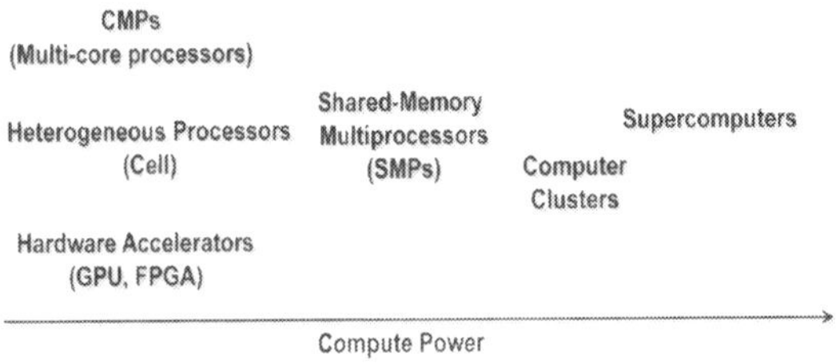

Fig. 1.1 Parallel compute hardware platforms.

compute node of present day clusters and supercomputers may include one or several multicore processors and even attached accelerators such as GPUs. As such, leveraging the ubiquitously available parallel compute hardware is a natural step for addressing the computational challenges of large-scale circuit simulation.

It is noteworthy that modern parallel processor architectures have the potential to enable circuit simulation approaches that are more massively parallel than explored ever before. For instance, on-chip interconnect networks in multicore processors are significantly advantageous over the off-chip interconnects in conventional multichip SMP (symmetrical multiprocessing) processors. On-chip interconnects span smaller distances and hence incur far less latency and power consumption. On-chip buses can be made much wider than those on a PCB. As a result, core-to-core and core-to-memory bandwidths and latencies in multicore processors are orders of magnitude better than those of SMPs, enabling finer grained parallelism. Computer clusters consisting of interconnected multicore processors can provide a greater amount of compute power, leveraged based upon a combination of chip-multiprocessing and distributed processing.

Moreover, the emergence of modern commodity heterogeneous platforms, comprising homogenous multicore microprocessors with attached accelerators (e.g., GPUs) or "fused" heterogeneous cores on the same die [2, 86], brings appealing new opportunities to the computing landscape. Impressive speedups may be gained if the workload is optimally partitioned, and the resulting partitioned tasks are efficiently processed on distinct (e.g., general-purpose vs. SIMD) processing cores that match each task's characteristics.

However, challenges still exist as we attempt to leverage this diverse spectrum of modern processor platforms for parallel circuit simulation. One central challenge, which is possibly long lasting, lies in the difficulty of scaling parallel simulation performance up to large numbers of processors due to the existence of unavoidable parallel overheads. To this end, it is essential to explore a plethora of new algorithmic and implementation level innovations that identify and exploit rich and orthogonal forms of parallelism in circuit simulation. It is desirable to exploit diverse (e.g., both task and data) parallelism with varying parallel granularity, synchronization overhead, and data movement patterns to fully utilize the available parallel compute power, and/or determine the best use of a particular type of processing units of a heterogeneous platform. The above considerations also underscore the important roles played by hardware characteristics (e.g., number and types of processors, cache size, memory bus bandwidth, and latency) and prompt algorithm/hardware interactions during the process of parallel algorithm design.

Looking in a different angle, we shall note that circuit properties have a definitive influence on the performance of a given simulation algorithm. In many ways, today's IC designs look different from older designs that were targeted during the early developmental phase of parallel simulation technologies several decades ago. In scaled fabrication processes, MOS transistors possess many more non-ideal device effects that must be included in transistor models and captured in simulation. We also see an explosion of design complexity, which is reflected not only by escalating device count but also by significantly increased parasitic coupling. Both factors stress the existing circuit simulation methodologies by creating large, complex, and highly coupled simulation instances. The desire toward accurate full-chip verification challenges the existing simulation technologies at yet another higher level, where a mixture of digital, analog, and memory blocks must be feasibly simulated.

While it is anticipated that "going for parallel" may offer an outstanding opportunity to tackle the above simulation challenges, it is observed that the ongoing technology and design trends clearly steer simulation technologies toward delivering higher performance, improved robustness, and accuracy. Nevertheless, these trends have broken the underlying assumptions of some classical parallel simulation techniques, rendering them less attractive or inapplicable for modern chip designs. This, however, strongly motivates the development of new parallel simulation approaches that are optimized for current IC designs in order to meet the desired performance, robustness, and accuracy requirements.

1.3 Focus and Organization of This Article

Circuit simulation is a broad field. Over the years, a myriad of simulation approaches have emerged, delivering solutions for a wide variety of digital, analog, mixed-signal, RF, and memory circuits as well as full-chip simulation needs with varying performance and accuracy targets. Clearly, examining all this large body of work from a parallel perspective is simply infeasible, given the scope limitation of this monograph.

As such, the main focus of this monograph is placed upon parallel transistor-level simulation techniques aimed at delivering full-SPICE accuracy. A comprehensive review, which is not meant to be exhaustive, is devoted to significant early developments in the past decades, and more recent work that has been stimulated by the arrival of the multicore computing era.

For the sake of completeness, key concepts behind fast-SPICE are succinctly reviewed to provide a basic exposure to the accuracy-performance tradeoffs made in this line of simulation techniques, which present a significant ongoing industrial development [1, 26, 32]. Elements of fast-SPICE techniques are contrasted with some of historical parallel simulation techniques.

The rest of this monograph is organized as follows. In Section 2, to set up a stage for this review, existing parallel circuit simulation works are classified according to two different views, *algorithmic level of parallel processing* and the *domain of parallel processing*. The two views offer a systematic examination of the types of parallelism that may be exploited in circuit simulation. Parallel direct simulation methods and the implementation issues are discussed in Section 3. Section 4 is devoted to two special classes of domain decomposition-based parallel simulation methods, nonlinear relaxation methods and waveform relaxation, which were the subjects of many historical research efforts. Other linear and nonlinear domain decomposition methods are presented in Sections 5 and 6, respectively. Some recent efforts in exploiting advanced numerical integration techniques for parallel simulation are described in Section 7. A multialgorithm parallel simulation framework, leveraging coarse-grained interalgorithm parallelism, is presented in Section 8. A succinct review of fast-SPICE is given in Section 9.

For most part of the aforementioned sections, the assumed underlying hardware architecture is general-purpose MIMD machine (e.g., multicore processors or computer clusters), which cover the bulk of previous and ongoing parallel simulation work. Some of the reviewed historical work was conducted on traditional vector machines.

Recent research activities on leveraging modern hardware accelerators and heterogeneous processors are reviewed in Section 10. Finally, confusions are drawn in Section 11.

2

Classifications of Parallel Simulation Methods

To put everything under a context, in this section, the formulation of circuit simulation problem is introduced. A variety of parallel simulation algorithms are then classified into two different perspectives, *algorithmic level of parallel processing* and the *domain of parallel processing*. The two classifications provide a basis for understanding the types of parallelism that exist in circuit simulation and the ways in which they may be exploited. Finally, a general discussion on algorithm and architecture interaction, an important issue, is provided.

2.1 Basic Differential–Algebraic Equation View of Circuit Simulation

In the time domain, an electronic circuit can be described using the following general form of differential–algebraic equations (DAEs) [74]

$$\frac{d}{dt}q(x(t),u(t)) + f(x(t),u(t)) = 0, \tag{2.1}$$

where $x(t) \in \mathbb{R}^N$ is the vector of unknown nodal voltages and branch currents, $u(t) \in \mathbb{R}^N$ is the input excitation, and $f(x(t),u(t))$, $q(x(t),u(t)) \in \mathbb{R}^N$ are the vector of resistive currents and the vector of charges or fluxes, respectively.

The above nonlinear system of equations is usually solved from a given initial condition say, $x(0) = x_0$ from time 0. Often the time, the DC circuit solution is solved to obtain x_0, where all the dynamic circuit elements are removed from the circuit and a DC input excitation is applied

$$f(x(0), u(0)) = 0. \tag{2.2}$$

To solve Equation (2.2) numerically, a nonlinear iterative method such as the Newton's method may be applied. At each Newton iteration, a linearized problem is solved to update the solution

$$G_k x_0^k = G_k x_0^{k-1} - f(x_0^{k-1}, u(0)), \tag{2.3}$$

where x_0^k and x_0^{k-1} are solution guesses at the k-th and $(k-1)$-th iterations, respectively, and $G_k \in \mathbb{R}^{N \times N}$ is the Jacobian matrix given by $G_k = \frac{\partial f}{\partial x_0}\big|_{x_0 = x_0^{k-1}}$.

Starting from the initial condition, a numerical integration method is used to discretize the DAE system in Equation (2.1) into a set of nonlinear algebraic equations. For instance, if the backward Euler method is applied, the following implicit nonlinear algebraic equation is solved to obtain the circuit response at each time point t_i

$$\frac{q(x(t_i), u(t_i)) - q(x(t_{i-1}), u(t_{i-1}))}{h_i} + f(x(t_i), u(t_i)) = 0, \tag{2.4}$$

where h_i is the discretization stepsize. From an electrical point of view, the system in Equation (2.4) corresponds to an equivalent DC circuit solution problem. In other words, by adopting a numerical integration formula, the time-domain transient response of the circuit is obtained by solving a sequence of equivalent nonlinear DC problems sequentially at all time points. The algorithm flow for nonlinear transient analysis is shown in Figure 2.1.

It shall be noted that some form of simulation accuracy and efficiency control is always desirable in practice. For transient analysis, the standard practice for doing so is to dynamically adjust the stepsize based on an error estimation scheme such as local truncation error (LTE) [74, 87]. The stepsize may be reduced if the estimated LTE is greater than a user-specified tolerance or reduced otherwise.

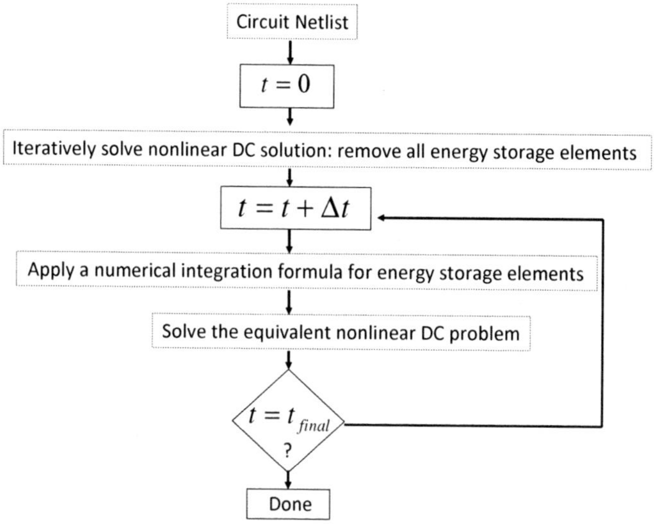

Fig. 2.1 Time-domain nonlinear transient analysis.

2.2 Other Problem Formulations and Elements of Circuit Simulation

The previous section touches on the basic DAE view of the circuit simulation problem and provides a feel of the numerical machinery behind the classical transient analysis. Indeed, in addition to transient analysis, there exist a variety of other circuit analyses which may be better formulated differently. For instance, for many analog and RF circuits employed in communication systems, specialized quasiperiodic state-steady analysis can be formulated and solved more efficiently in the frequency domain, the time domain, or a combination of both [63].

It is certain that some basic numerical machinery (e.g., matrix solution methods or nonlinear iterative methods), which forms the backbone of a circuit simulator, is always needed for any type of circuit analysis. Along this line, sophisticated numerical techniques such as advanced numerical integration methods, direct, and iterative methods have been developed.

However, a circuit simulator is much more than a "simple" DAE solver. Industrial-strength circuit simulators are built upon and must interface with sophisticated device models (e.g., models for MOSFETS,

BJTs, diodes) and complex parasitic networks that are used to faithfully model critical physical and electrical effects. Circuit netlists along with device models must be represented and stored in proper data structures for efficient processing during the phase of simulation. In addition to accurate device models, modern mixed-signal circuit simulators can support Verilog, VHDL, SPICE, Verilog-A, and Verilog-AMS models to facilitate design abstraction. Large parasitic networks may be reduced to accelerate simulation. A myriad of techniques including numerical, circuit and graph level techniques, parallelism, and software engineering has been used to deliver the desired levels of simulation capacity and accuracy (cf. [1, 14, 26, 32, 110]).

While the key focus of this monograph is on parallel and numerical simulation techniques, it is meaningful to assess these surveyed techniques under a complete context of circuit simulation.

2.3 Classifications of Parallel Simulation Approaches

Before getting started in developing a parallel circuit simulation algorithm, one has to think about what and how parallelism shall be explored. Fortunately, many opportunities exist for exploiting parallelism in a circuit simulation flow. A myriad of parallel simulation approaches has emerged to explore parallelism in various forms in the last several decades. Some of these opportunities are immediately visible within a standard simulation flow, while others may be only enjoyed through algorithmic innovations.

While a circuit simulation problem may be formulated in various ways, the standard nonlinear transient analysis flow does help shed some light on possible avenues for parallelizing circuit simulation. In the following subsections, we present two different ways for classifying existing parallel circuit simulation methods.

2.3.1 Classification by the Algorithmic Level of Parallel Processing

The first way to classify a parallel simulation algorithm is to examine the level at which parallelism is explored within the algorithm. As depicted in Figure 2.2, in this view, parallel methods are further

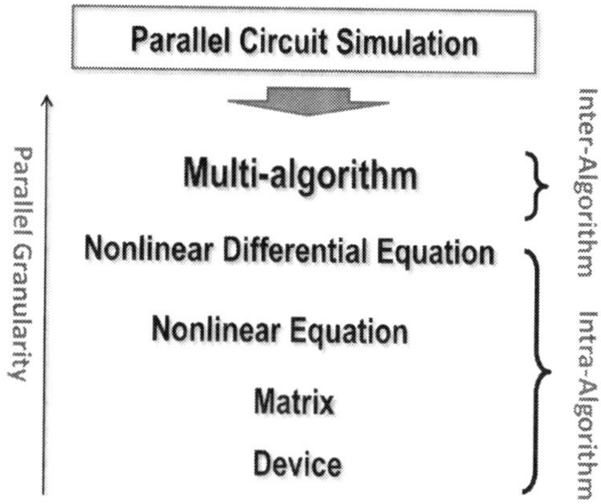

Fig. 2.2 Classifying parallel simulation algorithms by the algorithmic level of parallel processing.

classified into two categories: one that explores *intraalgorithm* parallelism and the other that explores *interalgorithm* parallelism.

Approaches that fall into the first category are "typical" parallel methods in the sense that they are a single simulation algorithmic entity within which various subcomponents of the algorithm are processed in parallel. With an increasing granularity of parallel processing, intraalgorithm parallelism can be explored at the levels of devices, matrices, (algebraic) nonlinear equations, or nonlinear differential equations (more precisely nonlinear DAEs), as shown in Figure 2.2. It is useful to note that these subcategories are not necessarily mutually exclusive; many real-life simulation implementations perform parallel processing at multiple algorithmic levels.

Device-level parallelism is one of the most widely adopted ways to parallelize circuit simulation. Complex device characteristics such as those of MOS transistors, BJTs, and diodes are described using sophisticated device equations, or device models (e.g., BISM3/3 models). Device models need to be evaluated many times in the iterative loop of a simulation algorithm, and hence are important sources of computational cost. For small-to-medium-sized circuits, model evaluation may take 70% of the total simulation time. Fortunately, multiple device model evaluations can be processed in parallel in a relatively straightforward way due to the apparent data parallelism [13, 43, 98, 112].

It is evident from Equation (2.3) that for transient analysis a linear system of equations must be solved at each Newton iteration. Solution of linear systems of equations is at the core of literally any simulation method. The cost of solving linear systems is dominant for large circuits. Thus, it is natural to parallelize this portion of simulation by using a parallel direct method [16, 98] or iterative method [23, 104, 106].

Parallelism can be also explored at the nonlinear equation level, for example, by using a nonlinear relaxation method to simultaneously solve scalar nonlinear equations associated with individual circuit nodes [18, 76, 113], or by applying the multilevel Newton method to update the unknown variables associated with multiple circuit blocks in parallel [117]. Modern fast-SPICE simulators often explore some forms of relaxation in order to avoid solving large systems of equations [92]. Relaxation can operate at the matrix or nonlinear equation level and is in principle amenable to parallelization.

Relaxation may be also applied in functional spaces or "waveforms" such that parallel processing is done at the differential equation level [65, 76, 98]. In this case, parallel computation of continuous time-domain responses of individual circuit nodes or blocks is performed. In a highly related manner, mixed-mode and fast-SPICE simulators (cf. e.g., [14, 26, 32]) employ multirate event-driven techniques, where circuit blocks are simulated using independent time-step control over active phases of operation.

Moving toward an even higher granularity, parallel multialgorithm simulation has been recently demonstrated [120, 121]. Different from more conventional single-algorithm-based parallel methods, in the multialgorithm approach, more than one simulation algorithms are launched to work on a single simulation task. Parallel speedups are achieved through on-the-fly cross-algorithm collaboration, a form of interalgorithm parallelism.

2.3.2 Classification by the Domain of Parallel Processing

The second classification provides a different view of the same set of parallel simulation methods and is complementary to the one that is just described. This classification is based upon the fact that intraalgorithm

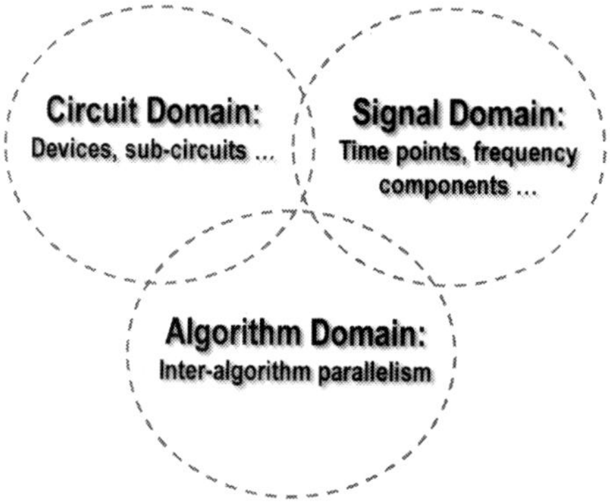

Fig. 2.3 Classifying parallel simulation algorithms by the domain of parallel processing.

parallelism is explored by partitioning a simulation task into smaller pieces so as to process these subtasks in parallel. Interalgorithm parallelism, on the other hand, is leveraged by concurrent execution of multiple algorithms on the same simulation task. Along this line, we perceive three possible "domains" of parallel processing, as shown in Figure 2.3.

A common way to create multiple subsimulation tasks is to perform circuit partitioning. As such, parallelism is explored in the *circuit domain*, where individual devices, circuit nodes, or subcircuits are processed concurrently. As an example, the approaches in [65, 76, 106, 117] fall into this category. Circuit-level partitioning is heavily exploited in fast-SPICE in order to achieve good simulation speed [92].

Parallel processing can be also done in the *signal domain*. For the time-domain transient analysis or other simulation problems that heavily rely on tracking the transient responses of the circuit, parallelism can be explored along the time axis. Namely, computations used to find the circuit responses at different time points may be processed in parallel [24, 98].

Similarly, for simulation methods that primarily operate in the frequency domain, the same may be said for the computational work associated with different frequency components [23]. Finally, the domain of parallel processing for multialgorithm simulation is the *algorithm domain*, in which algorithm diversity is explored to gain parallel speedups [120].

2.4 Algorithm and Architecture Interactions

It is worth noting that the central issue and also one of the key challenges in parallel processing is to achieve good parallel speedups up to a large number of processors on targeted architectures. To achieve this goal, it is often a necessity to create a rich set of (orthogonal) parallel opportunities so as to fully utilize the available parallel compute power. Therefore, exploring multiple forms of parallelism, such as ones that have been classified in the previous sections, is not only possible but also very desirable.

Some discussions on a highly relevant issue, the interactions between parallel simulation algorithms and the underlying hardware architecture, are in order. Algorithm and architecture interaction is a critical issue and can have a significant impact on the achievable parallel performance of an algorithm. The architectural characteristics of the processing platform shall be considered during the algorithm design phase when the selection of algorithms is made and algorithmic-level optimizations are performed. Furthermore, the parallel implementation of a chosen algorithm may be fine-tuned for targeted architectures. In principle, these two steps may be interleaved.

In general, algorithm and architecture interaction is a complex matter and covers a large space of research and practice. It may be algorithm, implementation, and machine specific. This is an area that deserves more future research and investigation. While recognizing the aforementioned factors, we briefly discuss several general considerations of this multifaceted matter for the case of parallel circuit simulation.

2.4.1 Architectural Perspective

Parallel computers such as the ones in Figure 1.1 can be classified according to the memory model used: shared, distributed, or attached (for Cell processor [86]) [68]. Another useful classification is based upon identifying the numbers of instruction and data streams: single-instruction-single-data (SISD, a.k.a sequential computers), single-instruction-multiple-data (SIMD), and multiple-instruction-multiple-data (MIMD) [33]. A particular type of parallel machines may be optimized for specific parallel workloads. For example, SIMD machines (e.g., GPUs and vector processors) are highly optimized for data-parallel applications but are not efficient for unstructured problems.

In addition to the above considerations, another important characteristic of parallel computers is that global and local memory references may take dramatically different amounts of time. In addition, the ratio between the two varies greatly across different machines. This ratio is around 100 for current multicore processors (e.g., an L1 cache hit takes one clock cycle and a cache miss takes about 100 cycles), while it may get closer to 10,000 for supercomputers [68]. The global reference latency relative to that of the local reference is an important factor to be considered when distributing data in a parallel program for which the locality of storage should be maximized.

As a key source of parallel overhead, communication and synchronization has a large impact on the performance of a parallel program. The current CMPs (i.e., multicore processors) have significantly improved core-to-core and core-to-memory bandwidths and latencies compared to other platforms such as SMPs and clusters, allowing for exploration of finer grained parallelism.

Important architectural characteristics including ones that are discussed above must be carefully considered while designing a parallel algorithm. Otherwise, unrealistic prediction of parallel performances may be resulted, which can mislead algorithm design.

2.4.2 Algorithmic Perspective

From an application point of view, to a large extent, the parallel granularity of an algorithm and the resulting dependencies (i.e., interactions between tasks) determine the performance of the algorithm on a hardware architecture. If a task is processed by a single process or thread, greater granularity helps increase the computation to overhead ratio, boosting the performance provided that all processing units are kept busy. On the other hand, with an increasing granularity, there tends to be a smaller number of independent tasks. At the point where the number of tasks becomes less than the number of available processes or threads, multiple of them have to be assigned to a single task. Then, it is important to be able to efficiently parallelize the work associated with a single task.

As classified in Figure 2.2, parallel simulation approaches operate with a wide range of granularity, and hence have varying degrees of match to specific architectures.

Device model evaluation represents a fair amount of computation and is independent among different devices, making it an ideal target for parallelization on standard parallel machines. The inherent data parallelism nature of device model evaluation has also been exploited on SIMD architectures (e.g., GPUs) [43, 55], heterogeneous architectures (e.g., Cell processor) [55], and FPGAs [53, 55], yielding good parallel speedups.

Compared with device model evaluation, parallelizing direct sparse matrix solving methods presents a much greater challenge [16, 44]. Architectural differences in factors such as the communication and synchronization overhead play an important role in determining the optimal data distribution and the resulting parallel speedups. For example, it is more desirable to use a static data distribution scheme for distributed platforms as dynamic redistribution of data may incur a significant cost. On MIMD machines, the sparsity of direct matrix solution methods offers opportunities for exploiting independencies between rows and columns, whereas the main challenge is to efficiently process unstructured operations with load balancing while minimizing parallel overheads. For matrices arising from circuit simulation, existing direct parallel matrix solvers provide a good parallel efficiency only up to a moderate number of processors on MIMD shared-memory and distributed machines. Iterative methods [23, 104, 106] may be developed in order to exploit a greater degree of parallelism. Sparsity creates significant challenges on SIMD machines (e.g., GPUs) as it creates irregular flow control and data access. Till now, little progress has been made on the development of general-purpose sparse matrix solution methods on SIMD platforms except for the case of structured circuit analysis problems.

To explore parallelism beyond what is offered by parallel device model evaluation and matrix solution, high-level simulation algorithms can be targeted. This is a broad space and many possibilities exist. The success of each approach, however, may be a complex function of fundamental (machine independent) algorithmic strength, circuit properties, and hardware architectural characteristics. Some general comments are provided below.

Parallel tasks with varying levels of granularity and dependency can be created in different manners. Circuit partitioning is a common way to create parallel tasks (e.g., [65, 76, 92, 106, 117]). The solution of the complete circuit can be computed via relaxation or some forms of domain decomposition [18, 76, 113, 117]. Parallelism may be also extracted out temporally, for example, by simultaneously computing the circuit responses at several time points [24, 98], at the waveform level [76, 65, 98], or via multirate event-driven simulation (cf. e.g., [26, 14, 32]). Parallelism with an even greater granularity can be made use of through parallel multialgorithm simulation [120, 121].

Along these lines, there exist delicate interactions between the algorithm design and the hardware architecture on which the algorithm runs. For instance, a relaxation method that seemingly creates many parallel tasks may run unwieldily slow on machines that have high interthread/process communication and synchronization overhead. This would be the case if the relaxation scheme requires many iterations to converge. This fact may effectively disqualify the approach for any efficient implementation on distributed-memory machines. In general, the use of coarse-grained parallelism tends to be less insensitive to the hardware architecture and may be efficiently implemented on different platforms (e.g., CMPs, shared-memory, and distributed-memory machines). Implementation strategies can also influence the performance of a parallel algorithm. For example, data locality produces better cache fitting, minimizes cache misses, and hence boosts the performance.

To penetrate the architectural barrier of SIMD processors for practical parallelization of circuit simulation tasks other than device model evaluation, more fundamental algorithmic-level innovations are needed. The recent work of GPU-accelerated power grid simulation is a good example [28]. A smart iterative algorithm is developed to expose the "hidden data-parallelism" in the application, making it possible to accelerate this big chunk of computation very efficiently on GPUs (more details in Section 10).

In general, tradeoffs involving granularity, dependency, overhead, and load balancing shall be properly made for a given hardware platform. Parallel performance modeling frameworks can be developed to capture key architecture-specific characteristics (e.g., cache/memory contention on shared-memory machines) and provide useful guidance for simulation algorithm design and tuning [122, 123]. The availability of high-performance computer clusters comprising interconnected homogenous, heterogeneous multi-/many-core processors and accelerators provides new opportunities and challenges in a broadened space. A holistic understanding of algorithm and architecture interaction is needed for exploiting a multiplicity of parallelism across various types of processing units. This is an area in which much more work remains to be done.

In the following sections of this monograph, a number of parallel simulation techniques will be discussed in detail. The discussion of these techniques will not strictly follow the order of the foregoing classifications. The sequence of the discussion is partially determined while considering the historical developments of such methods and their shared technical origins.

3

Parallel Direct Methods

Parallel direct methods refer to approaches that parallelize the direct simulation technique as implemented in the standard SPICE simulator [74]. Among other things, these methods employ a direct solution method for solving the underlying linear systems arising from a circuit simulation problem formulation.

From the analysis flow outlined in Section 2.1, the following major execution steps are identified for the nonlinear transient analysis

- Circuit parsing and simulation setup;
- Device model evaluation and matrix/RHS load for transistors;
- Matrix/RHS load for linear circuit elements;
- Sparse matrix solution;
- Error estimate and timestep control; and
- Output functions.

Out of these steps, a significant amount of runtime is spent on transistor model evaluation, matrix/RHS load, and sparse matrix solution. It shall be noted that in modern simulators parsing and other simulation setups can consume large amounts of CPU time. In practice, it may be meaningful to speed up this part of processing via parallelization too. However, the nature of problems and the techniques involved could be rather different. In the following discussion, only the first three computational steps are focused on.

While the parallelization of transistor model evaluation, matrix/ RHS load, and direct matrix solution appear to be conceptually straightforward, there exist several key algorithmic and implementation level considerations that are worth discussing. Furthermore, the direct methods as defined such have strengths in terms of robustness, accuracy, and convergence properties over many circuit types. Due to these reasons, parallel direct methods were the sole subject of several academic and industrial studies [7, 12, 13, 58, 125]. It is also useful to note that the basic ingredients of direct methods form the backbone of parallel circuit simulation. More sophisticated high-level parallel simulation approaches almost always integrate some of the key elements of parallel direct methods [24, 98, 106, 118, 120].

3.1 Parallel Device Model Evaluation and Matrix Load

Here, parallelism is exploited at the device/matrix level in the circuit domain according to the classifications presented in Section 2.3. Parallel transistor model evaluation can be implemented with minimal effort due to its data-independent nature. In comparison, several dedicate implementation level considerations exist for parallel matrix (RHS) load. The accumulation of the Jacobian matrix and the RHS requires summation of contributions from all devices. As an element in either the matrix or the RHS may have contributions from more than one device, such form of synchronization is needed.

Several parallel implementation choices exist for matrix (RHS) load with different tradeoffs in terms of efficiency, memory usage, and parallel implementation effort. These options have been explored in some early parallel circuit simulation work [98]. For example, the contributions of each device evaluation can be accumulated in a local memory storage first. Once the local accumulation is completed, the single lock to the entire matrix is acquired, and all the updates associated with the device evaluation are committed in one locking-and-releasing phase.

The likelihood of lock contention can be further reduced by using multiple locks to protect different rows or blocks of the matrix. It is also possible to completely separate the model evaluation and matrix accumulation; each of which can be parallelized in two separate phases [98]. For the above choices, load imbalance incurred by unequal assignment of device evaluation work among processors shall be minimized.

3.2 Parallel Direct Sparse Matrix Solution

Sparse matrix solution is a major runtime contributor to circuit simulation and can be the dominant source of the computational cost for large circuits. Therefore, it is meaningful to pursue parallelism at the matrix level.

3.2.1 Dedicated Parallel Direct Sparse Matrix Solution Methods for Circuit Simulation

Historically, in the community of circuit simulation, there have been many research developments dedicated to custom parallel spare matrix solvers specifically for circuit simulation [7, 12, 13, 98, 125]. Several parallel strategies with varying granularity exist for parallelizing sparse LU factorization. Loops inherent in the LU factorization may be parallelized with different granularity [13, 98]. Fine-grained parallelism can be explored by parallelizing element-based optimizations. Coarser grained parallel processing can be applied to the work associated with different rows or pivots.

Nevertheless, the approaches discussed above can only result in modest parallel gains on a small number of processors due to the inherent synchronization cost. A more promising approach is to leverage the inherent design hierarchy of the circuit under analysis by using node tearing, or mathematically, Schur complement [12, 13, 98]. Intuitively, for highly structured designs, it is possible to partition the circuit into a set of subcircuits (or domains) such that there is no coupling between these subcircuits while the subcircuits may connect to a small number of common interface nodes (interconnects). Such circuit partitioning puts the system matrix into the (nested) bordered block diagonal (BBD) form, where each subcircuit or the set of interface nodes corresponds to one diagonal matrix block while interactions between the subcircuits and the interface nodes are reflected in off-diagonal matrix blocks. To parallelize the matrix solution, each subcircuit block is first factorized in parallel. Subsequently, the linear system that corresponds to the interface is factorized, and a number of backward and forward substitutions are performed in parallel. A more detailed mathematical discussion of the process is provided in Section 5.2.

The efficiency of the above approach depends on several factors. The quality of circuit partitioning is important. The number of the interface nodes should be kept low so as to minimize the serial bottleneck of the second step of the matrix solution. However, this is conditional upon the characteristics of the design and the partitioning strategy adopted. Difference in the size of the subcircuits leads to load imbalance. This becomes a performance limitation particularly when the number of subcircuits is comparable to the number of processors. The problem can be alleviated by partitioning the circuit into a relatively large number of subcircuits. However, the increase in the size of the interface may become a concern.

3.2.2 Recent Progress in General-Purpose Parallel Direct Matrix Solvers

Instead of building a custom parallel matrix solver for circuit simulation, on-the-shelf general-purpose matrix solvers developed in the numerical analysis community may be adopted [16, 44]. A brief review on some recent progress on general-purpose parallel direct methods is provided below. More details may be found in [44].

A direct matrix solution method usually involves four steps: ordering, symbolic factorization, (numerical) factorization, and triangular solves. Each of these steps provides unique challenges and opportunities for parallelization.

In the first step, ordering heuristics are applied to a graph-based view of the matrix so as to order the matrix to yield certain properties. The nonzero pattern of the factorized matrix is computed in the second step, symbolic factorization. Knowing the non-zero pattern is beneficial for parallel processing as it provides a basis for distributing the data and computation among processing units. The ordering and symbolic factorization are also referred to as the *analysis* or *preprocessing* step. Typically, the two steps are not highly scalable in a parallel implementation, making it desirable to amortize the cost of the analysis step over the factorizations of multiple matrices with different numerical entries but the same non-zero structure.

The third step (numerical) factorization is often the most time and memory consuming step. There exist several difficulties and opportunities in parallelizing it due to the sparse nature of the matrix. The majority of difficulties arise from the fact that unstructured sparse operations must be managed with load balance during which parallel overheads associated with small numbers of row or column-based operations are well controlled. On the other hand, sparsity also offers new parallel opportunities in the sense that multiple columns of the factors may be computed simultaneously.

To explore such opportunities, a task dependency graph is often used to systematically express the dependencies in sparse matrix factorizations. A task dependency graph is a directed acyclic graph (DAG) with vertices corresponding to tasks associated with different rows and columns, and the edges representing intertask dependencies. A task may be processed only if all of its predecessors have been processed. In the case of symmetrical matrices, task dependency graphs are also called elimination trees. One example of elimination tree is shown in Figure 3.1. Based upon a task dependency graph, which is in fact often constructed in the symbolic factorization step, the parallelization of the numerical factorization and the triangle solves becomes a problem of scheduling tasks onto different processes or threads. It shall be noted that the ordering step has a significant impact on the resulting task dependency task. One of the best known ordering heuristics, nested dissection, may help produce short and wide dependency graphs that lead to a high degree of parallelism.

In practice, supernodes, instead of single rows or columns, may be explored in conjunction with a multifrontal method, which is often the formulation of choice for high-performance parallel implementations [44]. A supernode contains multiple consecutive rows (columns) that share the same nonzero pattern. The use of supernodes leads

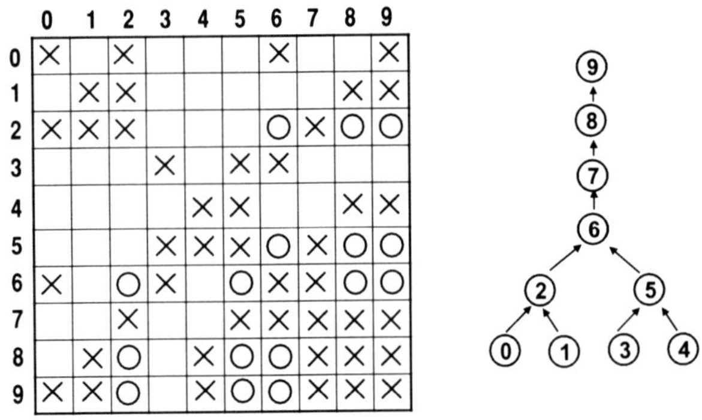

Fig. 3.1 A symmetric sparse matrix (left), and its elimination tree (right). The original nonzero entries are labeled by X's and the fill-ins by circles.

to a compressed graph and the ordering heuristics can be more efficiently applied to compute a task dependency graph in which vertices are supernodes. Moreover, within a supernode, operations can be performed more efficiently on a dense submatrix. Supernodes also increase the granularity of the parallel tasks, helping to minimize the relative contribution of the parallel overhead.

In a multifrontal method, a supernode and its update contribution are stored in small dense matrices called frontal matrices. The paradigm of multifrontal methods is desirable since a large portion of factorization work is performed in dense frontal matrices where a strong data locality exists. In addition, updates are accumulated and propagated from the leaf supernodes to the root of the dependency graph efficiently to avoid excessive communication overhead. Both data parallelism and task parallelism can be explored in multifrontal methods. A large number of leaf nodes are processed in a data-parallel fashion starting from the bottom of the tree while multiple processes or threads may be employed to process parallel dense matrix operations for a supernode toward the root where the number of remaining tasks becomes diminishingly smaller.

The fourth step, triangular solves, is based upon the factors that are computed in the preceding factorization step. As a result, the parallelization of the triangular solves must use the same data distribution employed in the parallel factorization step. Hence, an efficient parallel factorization, in particular, the data distribution scheme, as described before, is also critical for the following triangular solves.

The use of the state-of-the-art parallel direct matrix methods has been shown to be a critical enabler for many circuit simulators. However, one of the limitations of the existing parallel direct matrix solvers is that they tend to offer high parallel efficiencies only up to a moderate number (e.g., a couple of tens) of processors. To provide a greater parallel speedup, particularly for large-scale circuit simulation, iterative methods or other more sophisticated simulation approaches need to be developed as detailed in the following sections.

4

Parallel Relaxation Methods

Relaxation-based methods were developed for simulating digital MOS circuits in the time domain [76]. This was an active area of research that attracted a large amount of research effort a couple of decades ago. This is one of the motivations for revisiting this family of simulation techniques so as to provide a historical perspective on the development of parallel circuit simulation techniques.

Relaxation can be applied at the nonlinear equation level. In this case, the solutions to nonlinear systems of equations resulted from the discretization of the nonlinear DAE of the circuit are found by applying a nonlinear relaxation method such as Gauss–Jacobi or Gauss–Seidel method [76]. Relaxation can be also applied at the nonlinear differential equation level. This leads to the waveform relaxation (WR) method [5, 65, 118]. The application of relaxation leads to a decomposition of the simulation problem into smaller systems of nonlinear algebraic or differential equations that can be solved in parallel. For the sake of convenience of reference, we refer to the Gauss–Jacobi or Gauss–Seidel methods that are applied at the nonlinear equation level as nonlinear relaxation methods. We adopt this terminology despite the fact that the relaxed simulation problem resulted from WR can be also solved in a Gauss–Jacobi or Gauss–Seidel fashion.

The relaxation methods lend themselves to parallelization [5, 18, 47, 48, 51, 52, 80, 82, 90, 98, 113, 115, 118, 124]. According to the classifications presented in Section 2.3, the domain of parallel processing in which the nonlinear relaxation methods operate is the circuit domain, and the algorithmic level at which parallelism is explored is the nonlinear equation level. Since waveform relaxation extends the relaxation to time-domain waveforms, its parallel processing domain is the combination of the circuit and time domains, and the algorithmic level of parallel processing is the nonlinear differential equation level. For a class of digital MOS circuits, relaxation methods were shown to provide significant speedups for transient analysis [76].

Historically, the application of these relaxation methods relied on several characteristics of digital MOS circuits, particularly, unidirectionality. For instance, one observation is that if the gate-to-drain and gate-to-source capacitances are small the gate of a MOSFET transistor may be considered as disconnected from both the drain and the source. Under this assumption, one may speculate that signals mostly propagate from the inputs toward the outputs in unidirection. This physical property may be explored to identify relatively loosely coupled subsystems either in the space or the time to apply relaxation.

The above assumptions are not necessarily true for present-day manufacturing technologies and designs as will be discussed later in this section. Nevertheless, the historical work on the subject provides a useful context for understanding the general concept of relaxation and its potential application to parallel processing. Indeed, ingredients of the ideas reviewed in this section have evolved into the current industrial practice of fast-SPICE technology, which will be discussed in Section 9.

4.1 Necessary Conditions for the Existence of a Unique Circuit Solution

Before discussing any specific relaxation methods, we first consider the conditions these relaxation methods typically assume to guarantee for the existence of a unique circuit solution [118].

We rewrite Equation (2.1) in a different form

$$\frac{\partial q(x(t),u(t))}{\partial x}\dot{x}(t) = -\frac{\partial q(x(t),u(t))}{\partial u}\dot{u}(t) - f(x(t),u(t)). \tag{4.1}$$

Define $C(x(t),u(t))$ by $\frac{\partial q(x(t),u(t))}{\partial x}$ and $b(x(t),u(t))$ by $-\frac{\partial q(x(t),u(t))}{\partial u}\dot{u}(t) - f(x(t),u(t))$. Now we have the following new form of Equation (4.1) under a given initial condition x_0

$$C(x(t),u(t))\dot{x}(t) = b(x(t),u(t)), \; x(0) = x_0. \tag{4.2}$$

Lemma 4.1. [118] if $C(x,u)^{-1}$ exits for all x and u and is uniformly bounded with respect to x and u, $b(x,u)$ is Lipschitz continuous with respect to x for all u, $u(t)$ is piecewise continuous, then there exists a unique solution to Equation (4.2).

It is worth noting that the above Lemma provides a set of *necessary* conditions for Equation (2.1) to have a unique solution. Along with others, the above conditions are needed to guarantee the convergence of the relaxation methods as discussed as follows. It should also be noted that the necessary conditions presented here impose certain practical constraints on circuit topologies and elements, as discussed later in the section.

4.2 Nonlinear Relaxation Methods

Returning to Equation (2.2), but rewrite it slightly different as

$$W(x) = W(x(t_i)) = 0, \tag{4.3}$$

where $W(x(t_i)) = \frac{q(x(t_i),u(t_i)) - q(x(t_{i-1}),u(t_{i-1}))}{h_i} + f(x(t_i),u(t_i))$, and for convenience of notation $x(t_i) = [x_1(t_i), x_2(t_i), \ldots, x_N(t_i)]^T$ is denoted by $x = [x_1, x_2, \ldots, x_N]^T$. Note that Equation (4.3) is a set of N nonlinear equations, where $W(x) = [w_1(x), w_2(x), \ldots, w_N(x)]^T$ with $w_i : \mathbb{R}^N \to \mathbb{R}$.

If the Gauss–Jacobi relaxation method is applied to solve Equation (4.3), at each, e.g., the $(k+1)$-th, relaxation step, x_i is updated by solving the following scalar implicit nonlinear equation [81]

$$w_i(x_1^k, \ldots, x_{i-1}^k, x_i^{k+1}, x_{i+1}^k, x_N^k) = 0. \tag{4.4}$$

If the Gauss–Seidel method is used, the following equation is solved instead

$$w_i(x_1^{k+1},\ldots,x_{i-1}^{k+1},x_i^{k+1},x_{i+1}^k,x_N^k) = 0. \tag{4.5}$$

The Gauss–Jacobi relaxation-based time-domain transient analysis algorithm is outlined in Algorithm 1. The Gauss–Seidel-based methods can be obtained similarly. On top of these base-line algorithms, techniques that exploit latency such as the event-driven selective trace algorithm can be further introduced to speed up the simulation [76].

Algorithm 1: The Gauss–Jacobi Relaxation Method.

```
t_n = 0;
Solve for the DC circuit response x(t_n) = x_DC;
t_n ← t_{n+1};
while t_n < TMAX do
    x^0 ← x(t_{n-1});
    k ← 0;
    repeat
        for i ← 1 to N do
            Solve w_i(x_1^k, ..., x_{i-1}^k, x_i^{k+1}, x_{i+1}^k, x_N^k) = 0 for x_i^{k+1};
        end
        k ← k + 1;
    until ||x^k − x^{k-1}|| < ε;
    x(t_n) ← x^k;
    t_n ← t_{n+1};
end
```

A theoretically and practically important issue with any relaxation method is the property of convergence. It has been shown that under certain conditions both the nonlinear Gauss–Jacobi and Gauss–Seidel algorithms will converge for transient simulation of the system in Equation (2.1), as shown in the following theorem [76, 118].

Theorem 4.1. Along with the conditions specified in Lemma 4.1 that guarantees that Equation (2.1) has a unique solution, if $f(x,u)$ is continuously differentiable with respect to x uniformly in u,

$C(x(t), u(t)) = \frac{\partial q}{\partial x}(x, u)$ is strictly diagonally dominant uniformly overall for all x and u, and $x(t_{i-1})$ is set to be the initial guess for solving the unknown $x(t_i)$, the Gauss–Jacobi and Gauss–Seidel methods will converge for the numerical integration scheme of Equation (2.4) used to discretize Equation (2.1) for all $h_i < h_t$, where h_t is some positive number.

Note that the above theorem also holds for any standard consistent multistep method in addition to backward Euler of Equation (2.4). As part of the above theorem, one key condition needed to guarantee the convergence is that $C(x(t), u(t))$ is strictly diagonally dominant. To get a sense of this theoretical result, consider a nonlinear circuit that only contains linear capacitors that are represented by a fixed capacitance matrix C. In this case, Equation (2.4) reduces to

$$Cx(t_i) = Cx(t_{i-1}) + h_i f(x(t_i), u(t_i)) = 0. \tag{4.6}$$

Theorem 4.1 says that the nonlinear relaxation methods will converge if C is strictly diagonally dominant and h_i is smaller than some number. In the limit, as $h_i \leftarrow 0$, Equation (4.6) further reduces to

$$Cx(t_i) = Cx(t_{i-1}), \tag{4.7}$$

which is a set linear system of equations. Clearly, if C is strictly diagonally dominant, both the Gauss–Jacobi and Gauss–Seidel will converge.

4.3 Waveform Relaxation

Relaxation can be also applied in the domain of "waveforms". In the so-called waveform relaxation method [65], the Gauss–Jacobi or Gauss–Seidel-type relaxations are brought up to the the differential equation level for solving large digital MOS circuits.

Broadly speaking, waveform relaxation can be thought as a way to explore the multirate nature of the circuit being simulated. State variables in large systems may evolve with considerably different rates over the time. The application of standard numerical integration forces all state variables to be discretized using the same stepsize. This choice often renders the common stepsize small in order to trace any existence of fast activities in the entire circuit.

If each differential equation were solved independently of others, for example, by picking up different stepsizes or even a different numerical integration method, the multirate behavior of the circuit could be naturally explored. This is the motivation behind waveform relaxation [65, 118]. For the system of Equation (2.1), a baseline waveform relaxation algorithm is shown in Algorithm 2 using Gauss–Jacobi. The one that is based upon Gauss–Seidel can be similarly developed.

Algorithm 2: Gauss–Jacobi Waveform Relaxation Algorithm.

Provide an initial waveform solution $x^0(t)$, $t \in [0, TMAX]$ with a given initial condition $x^0(0) = x_0$;

$k \leftarrow 0$;

repeat

  $k \leftarrow k + 1$;

  for $i \leftarrow 1$ **to** N **do**

    Solve

$$\frac{d}{dt} q_i(x_1^{k-1}, \ldots, x_{i-1}^{k-1}, x_i^k, x_{i+1}^{k-1}, \ldots, x_N^{k-1}, u) + f_i(x_1^{k-1}, \ldots, x_i^{k-1}, x_i^k, x_{i+1}^{k-1}, \ldots, x_N^{k-1}, u) = 0$$

    for $x_i^k = x_i^k(t), x_i^k(0) = x_{0,i}, t \in [0, TMAX]$;

  end

until $||x^k - x^{k-1}|| < \epsilon$;

As for the relaxation methods applied at the nonlinear equation level, the convergence property of waveform relaxation is a critical issue. The following convergence result exists regarding the system of Equation (4.2) [118].

Theorem 4.2. Along with the conditions in Lemma 4.1, if $C(x, u)$ is strictly diagonally dominant uniformly overall for all x and u and Lipschitz continuous with respect to x for all u, and $x^0(t)$ is differentiable, then for any bounded interval $[0, T]$, the waveforms $\{x^k\}$ produced by the Gauss–Jacobi or Gauss–Seidel waveform relaxation algorithm will converge uniformly for the solution of system (Equation (4.2)).

4.4 Parallelization of Nonlinear Relaxation and Waveform Relaxation Methods

To put the discussion under an historical perspective, we report some early demonstration of effectiveness of parallel nonlinear relaxation and waveform relaxation methods. When interpreting these results, care needs to be taken as these experiments were performed on older generations of machines.

For large digital MOS circuits, the Gauss–Seidel-based iterative timing analysis (ITA) as implemented in the SPLICE program was shown to provide more than one order of magnitude speedups over the standard SPICE simulator (SPICE2G) [76], where selective trace algorithm was also adopted to explore event-driven simulation. SPLICE was parallelized on both the Digital VAX11-780 and the BBN Butterfly machines [18, 51]. Scheduling and queue contention overhead was identified as the main parallel performance barrier and hence minimized in the implementation. Good parallel speedups were demonstrated up to a large number (about 40) of processors.

Waveform relaxation and its parallelization have been focused on in many research projects. One implementation, RELAX2, has been shown to be able to provide more than one order of magnitude simulation speedups for digital circuits [76]. Speeding up ideas such as bypass and partial convergence were shown to be effective. Furthermore, splitting the simulation period into smaller time windows, partitioning circuits into strongly coupled clusters, and ordering the updates properly improved the convergence [76, 118]. On a shared-memory Sequent Balance 800 multiprocessor computer, a sevenfold parallel speedup was reported for simulating memory and microprocessor control circuits using RELAX2.

Relaxation-based methods such as ITA and waveform relaxation have also been implemented on an Alliant FX/8 cluster comprising eight vector processors [98]. The parallel ITA implementation of the PSPLICE3 program showed parallel speedups of 2.4 to 6.2 times using eight processors over the serial implementation of ITA in the SPLICE3.1 code [98]. The parallel implementation of waveform relaxation on the eight-processor Alliant FX/8 cluster demonstrated a parallel speedup of 1.4–5.9 times. The choice of Gauss–Seidel vs. Gauss–Jacobin as well as windowing did influence the parallel runtimes. Extensions to parallel simulation on multiple Alliant FX/8 clusters were demonstrated in [48]. High-level waveforms obtained by fast simulation techniques were explored before the start of the waveform relaxation simulation to guide circuit partitioning, window size selection, and task allocation in [115].

Parallel waveform relaxation of circuits containing global feedbacks were studied in [52]. More recently, parallel waveform relaxation has also been demonstrated for the simulation of large-scale coupled linear interconnect networks [5, 82].

4.5 Applicability to Modern Designs and Technologies

Robustness is a general issue for nondirect methods including the relaxation methods discussed above. While relaxation methods applied either at the equation level or the differential equation level yielded great success for simulating digital MOS circuits, particularly in the 1980s and 1990s, some characteristics of these relaxation methods do limit their application with varying degrees.

For instance, the aforementioned necessary conditions for Equation (2.1) to have a unique solution, which are also part of the conditions for the relaxation methods to converge, require that $C(x,u)^{-1}$ exists. Practically, this implies that there must be a grounded capacitance at each circuit node, which may not be a severe constraint, particularly for realistic extracted layouts with full parasitics. The same requirement imposes certain constraints on how voltage sources and inductors may be modeled.

Furthermore, one of the conditions stated in Theorem 1 for the Gauss–Jacobi and Gauss–Seidel methods to converge is that the step-size must be well controlled. The implications of this condition shall be carefully investigated in practice as it may force the simulator to take large numbers of small steps.

As stated in Theorems 4.1 and 4.2, one of the more demanding conditions to ensure the convergence of the Gauss–Jacobi/Gauss–Seidel and waveform relaxation methods is that capacitance matrix $C(x,u)$

is strictly diagonally dominant. This appears to be a stringent requirement for circuits designed for modern IC processes. This is due to the fact that parasitic couplings between circuit nodes can be rather significant in scaled technologies.

The relaxation-based methods implicitly exploit the unidirection of signal propagation to achieve fast convergence. However, this assumption may break down due to several reasons. Analog circuits tend to have several tightly coupled local and global feedbacks and hence the signal propagation is not necessarily dominant in one direction. The same may be said for modern digital MOS circuits for reasons such as the existence of strong Miller effects and the use of pass gates.

It may be fair to say that the above reasons may have at least partially contributed to the fact that relaxation-based methods are not widely used today except for cases where coupling between different circuit nodes or blocks can be well quantified [5, 82]. Nevertheless, useful ingredients of the nonlinear relaxation methods reviewed here have successfully evolved into the development of modern fast-SPICE type simulators, which is a significant ongoing industrial practice. A brief review of the fast-SPICE simulation technology is given in Section 9.

5

Parallel Domain Decomposition Methods I

Domain decomposition refers to a class of methods for solving linear or nonlinear systems of equations, primarily arising from the discretization of partial differential equations (PDEs) [103]. *Domain decomposition* may mean different things under various contexts. In a way that is mostly relevant to the focus of this monograph, a domain decomposition method finds the solution to a large system by subdividing it into smaller subproblems (i.e., defined on subdomains) and solving these subproblems separately. While the circuit analysis problems of interest of this monograph do not necessarily originate from a underlying continuous PDE formulation, the same mathematical machinery of domain decomposition may be still applicable.

Broadly speaking, Gauss–Jacobi, Gauss–Seidel, and waveform relaxation methods discussed earlier are special cases of domain decomposition methods. In this section, we will discuss domain decomposition methods for solving linear systems. In this regard, the direct matrix solution method using the bordered block diagonal (BBD) form mentioned in Section 3.2.1 is a domain decomposition method. It will be formally presented in this section.

It should be noted that the same class of the methods may be immediately applicable to nonlinear systems, by applying them to find solutions of linearized systems arising from the adoption of the Newton's method. Furthermore, domain decomposition that is based on solving nonlinear subdomain problems can be also developed. Both types of domain decomposition methods will be the subject of the next section.

5.1 Linear Circuit Analysis

As a special case of Equation (2.1), a linear circuit can be modeled by a linear differential equation

$$C\frac{d}{dt}x(t) + Gx(t) = bu(t), \tag{5.1}$$

where C is the capacitance matrix (albeit it may have contributions from inductive elements), G is the conductance matrix, and b is the input matrix (vector) that connects external input $u(t)$ to the system.

Applying a standard numerical integration method to Equation (5.1), in this case backward Euler, leads to a system of linear equations for each time point

$$C(x(t_i) - x(t_{i-1}))/h_i + Gx(t_i) = bu(t_i). \tag{5.2}$$

Define $A = C/h_i + G$ and $v = Cx(t_{i-1})/h_i + bu(t_i)$. Write Equation (5.2) in a slightly different form

$$Ax = v, \tag{5.3}$$

where for convenience of notation, the time argument in $x(t_i)$ is dropped.

In the remainder of this section, two domain decomposition methods will be discussed. According to the classifications presented in Section 2.3, the methods we will discuss in this section operate at the matrix level. Their domain of parallel processing is the circuit domain since submatrix problems created in these methods correspond to partitioned circuit blocks.

5.2 Parallel Schur Complement

Suppose that the linear system in Equation (5.3) is partitioned into $K+1$ nonoverlapping subdomains $D = \{D_1, D_2, \ldots, D_K, \Omega_{\text{int}}\}$. There is no coupling between subdomains $D_1, D_2, \ldots, D_K$ while each D_i may be coupled to Ω_{int}, which represents a set of interface nodes between D_i's. Conformingly, the unknown vector x is reordered and partitioned as: $x = [x_1^T, x_2^T, \ldots, x_K^T, y]^T$ with $x_i \in D_i$ and $y \in \Omega_{\text{int}}$. Now Equation (5.3) is written in the following form, which corresponds to a reordering of the unknown variables

$$\begin{bmatrix} A_1 & & & & F_1 \\ & A_2 & & & F_2 \\ & & \ddots & & \vdots \\ & & & A_K & F_K \\ E_1 & E_2 & \ldots & E_K & J \end{bmatrix} \begin{bmatrix} x_1 \\ x_2 \\ \vdots \\ x_K \\ y \end{bmatrix} = \begin{bmatrix} v_1 \\ v_2 \\ \vdots \\ v_K \\ z \end{bmatrix}. \tag{5.4}$$

The above matrix structure is known as the bordered block diagonal (BBD) form [12, 13, 98, 111].

The above system in the BBD form can be solved using Schur complement, the bulk of which may be parallelized as follows. For each subdomain except for the interface, we have:

$$x_i = A_i^{-1}(v_i - F_i y), \quad i = \{1, \ldots, K\}. \tag{5.5}$$

Substituting this expression into the last block row of Equation (5.4) gives

$$\left(J - \sum_{i=1}^{K} E_i A_i^{-1} F_i \right) y = z - \sum_{i=1}^{K} E_i A_i^{-1} v_i, \tag{5.6}$$

where matrix $J - \sum_{i=1}^{K} E_i A_i^{-1} F_i$ is the Schur complement of blocks $A_1, \ldots, A_K$. Vector y can be obtained by solving the linear system in Equation (5.6). Then, using the known y, each x_i is found by solving the linear system corresponding to Equation (5.5)

$$A_i x_i = v_i - F_i y, \quad i = \{1, \ldots, K\}. \tag{5.7}$$

The parallelization of the above process is straightforward. An LU factorization for each A_i is first obtained in parallel. Then, to form the system in Equation (5.6), each $A_i^{-1} F_i$ and $A_i^{-1} v_i$ in the left and right-hand sides is computed, again, in parallel by reusing the LU factor. Once formed, Equation (5.6) is solved usually in a serial process.

Finally, each x_i is found by solving the system in Equation (5.7) again in parallel by reusing the existing LU factor.

5.2.1 Parallel Implementation Considerations

It is easy to see that the above parallel Schur approach will be efficient if the cardinality of Ω, or the number of interface nodes, is small. This corresponds to a small linear system for the interface nodes in Equation (5.6), which is easy to solve. In this case, the rest of the computation, which dominates the overall computational cost, is fully parallelized. It is apparent that the parallel Schur approach is ideal for large structured designs where the system can be partitioned into loosely coupled blocks where interblock coupling can be confined in a small number of interface nodes. From a circuit design point of view, the above condition often implies that the system is highly structured due to the existence of design hierarchy.

However, in practice, the circuit under analysis may not have any clean partitioning boundary, for instance, in the case of unstructured mesh circuits. Under such a situation, finding the solution to Equation (5.6) becomes a severe bottleneck. In fact, not only solving Equation (5.6) may become expensive, forming it can be very costly too as each F_i may have many columns in it.

From the foregoing discussions, it is clear that the solution to the interface system of Equation (5.6) can become a severe serial bottleneck in practice. The size of the interface is determined by the quality of circuit partitioning. In addition to finding better partitioning strategies, to reduce the serial work associated with the interface, partial LU factorization and forward substitution may be applied [13]. The basic idea is to start partially factorizing and substituting Equation (5.6) immediately after a subcircuit has been processed and loaded into the interface matrix.

Circuit partitioning is also critical for controlling the granularity of parallel processing and the resulting load imbalance. It is unreasonable to expect that all the domains will be partitioned with identical size. Therefore, a large domain shall be further partitioned to reduce the granularity and produce a sufficiently large number of domains to guarantee high processor utilization. A two-step partitioning strategy is adopted in [13]. In the first step, closely coupled circuit elements are clustered to form primary functional blocks. Then, in the second step, a global bisection algorithm (i.e., Kernighan and Lin Algorithm [59] or Fiduccia and Mattheyses Algorithm [31]) is employed to repeatedly bisect the circuit to produce a desirable number of partitions with nearly equal size. It is noteworthy that other partitioning objectives can be also considered, for instance, minimization of amount of fill-in and factorization cost. An excellent more recent partitioning tool (*nested dissection*) to use is hMETIS [57].

5.3 Parallel Schwarz Methods

To resolve the bottlenecks associated with Schur complement, a Schwarz method may be adopted [103]. For Schwarz methods, the domain of the original linear system is again partitioned into a number of subdomains, say K subdomains: $D = \{D_1, D_2, \ldots, D_K\}$ while one difference is that now coupling between subdomains D_i's is allowed.

There are several types of Schwarz methods. In nonoverlapping Schwarz methods, subdomains must be disjoint while their overlapping counterparts allow for overlaps between different sub-domains. Schwarz methods can be also classified into the additive and multiplicative methods. The multiplicative Schwarz method processes different domains with an order, in a way similar to the block Gauss–Seidel scheme. In fact, the nonoverlapping additive Schwarz method is equivalent to the classical block-Jacobi method. Similarly, the nonoverlapping multiplicative Schwarz method is essentially the block-Seidel method.

5.3.1 Parallel Additive Schwarz Method

Consider the mesh-like linear circuit in Figure 5.1 (left). The spatial domain of the mesh is partitioned into several subdomains where there may exist overlaps between these domains. Correspondingly, the partitioning is reflected in the matrix of the formulated circuit analysis problem (Figure 5.1 (right)).

For each subdomain i, define a rectangular restriction matrix R_i such that $A_i = R_i A R_i^T$ corresponds to the rows and columns of

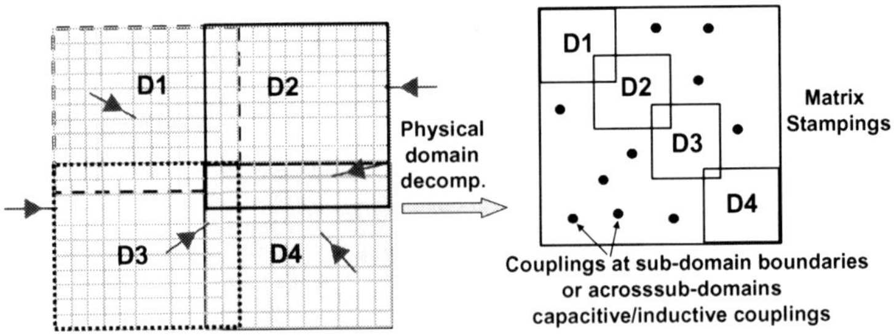

Fig. 5.1 A mesh circuit partitioned into overlapping subdomains (left), and the corresponding matrix structure.

subdomain i in A, which is the i-th diagonal block in the matrix. Essentially, R_i is a restriction operator that maps a vector from the full domain D to subdomain D_i. R_i^T, on the other hand, is a prolongation operator that maps a vector from subdomain D_i back to D. Starting from a given initial solution vector x^0, additive Schwarz method iteratively finds the solution to $Ax = v$ till the convergence is reached

$$x^{j+1} = x^j + \sum_{i=1}^{K} R_i^T A_i^{-1} R_i (v - Ax^j), \quad j = \{0, 1, 2, \ldots\}. \tag{5.8}$$

It is straightforward to see that the above additive Schwarz method can be readily parallelized by processing the computation associated with different subdomains concurrently. However, important tradeoffs shall be considered for practical implementation. Algorithmically, a higher degree of overlap between subdomains may lead to fast convergence of the additive Schwarz method. However, this comes with increased interthread/process communication cost. Therefore, rate of convergence and communication must be carefully traded off. In addition, the amount of overlap shall be determined while considering other factors such as the overall partitioning scheme and the resulting impact on load balancing.

5.3.2 Parallel Multiplicative Schwarz Methods

In the multiplicative Schwarz method, subdomains are processed in a particular order in a way similar to the block-Seidel method. Suppose that the processing of the subdomains follows the order of

$D_1, D_2, \ldots, D_K$. The multiplicative Schwarz method can be mathematically described by the following iterations

$$x^{j+1} = \prod_{i=1}^{K} (I - R_{K+1-i}^T A_{K+1-i}^{-1} R_{K+1-i} A) x^j + Pv, \quad j = \{0,1,2,\ldots\}, \tag{5.9}$$

where

$$P = \left[I - \prod_{i=1}^{K} (I - R_{K+1-i}^T A_{K+1-i}^{-1} R_{K+1-i} A) \right] A^{-1}.$$

The multiplicative Schwarz method is less amenable to parallel processing compared with its additive counterpart. Nevertheless, it is still possible to parallelize multiplicative Schwarz by using a coloring scheme albeit only disjoint subdomains groups can be processed in parallel.

In [126, 128], an efficient partitioning-based locality-driven parallel on-chip power grid analysis approach was presented. While the approach was originally conceived completely from a circuit point of view, we show its connection to the multiplicative Schwarz method.

The parallel approach of [126, 128] exploits the strong locality exhibited in designs with flip-chip (C4) type packages. In the power grid of such designs, there exist a large number of Vdd/ground pads that are connected to C4 bumps on the package. For the on-chip power grid analysis, these pads are considered as ideal Vdd/ground nodes. Since the Vdd/ground pads spread out in the power grid, the circuit response at a particular node is mostly influenced by the current loadings in the near neighborhood as well as by the nearby Vdd/ground pads which act as local current sinks.

As shown in Figure 5.2, the locality behavior of the on-chip power grid is exploited as follows. The entire power grid is partitioned into a set of nonoverlapping subdomains (or partitions) indicated by the black dashed lines. For each partition, the exact values for the boundary node voltages (or the currents entering the boundary from neighboring partitions) are considered as the boundary conditions for the partition. If the boundary condition is known, the voltage response within the partition can be solved independently of others using the known boundary condition in parallel.

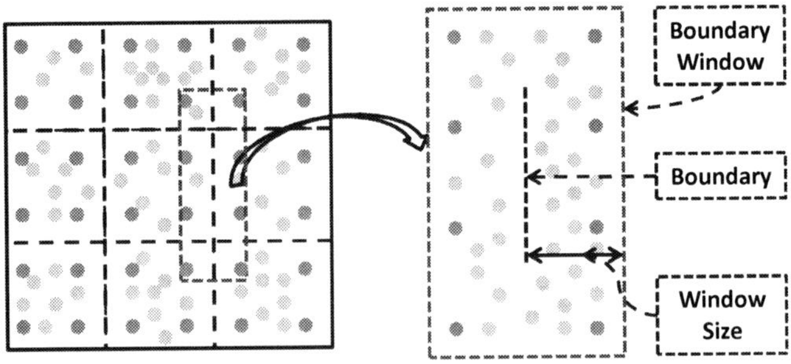

Fig. 5.2 Parallel locality-driven power grid analysis. Red dots represent C4 bumps. Green dots represent current drains. Black dashed lines represent partition boundaries. Blue dashed lines represent boundary windows (from [128] ©[2009] IEEE).

To find the boundary condition efficiently, a local boundary window that encloses a part of partitioning boundary at its center is considered (the blue dashed line in Figure 5.2). Due to the locality, it may be sufficient to solve a relatively small boundary window to find the circuit response at the center which is the desired boundary condition for the partitioning scheme. To make the approach accurate, each boundary window should enclose certain number of local Vdd/ground pads. To practically implement the idea, the size of the boundary window is progressively increased until the solved circuit response at the center of the window converges. To find all the needed boundary conditions, multiple windows are created and solved in parallel. It was shown in [128, 126] that the approach is rather efficient for analyzing large power grids. The same spirit was embraced under the context of optimization in [127] for parallel sizing optimization of on-chip power grids.

While the above approach was motivated by recognizing the electrical characteristics of power grids with a flip-chip package [126, 128], it can be also viewed mathematically as a variant of the overlapping multiplicative Schwarz method. In this perspective, the entire domain is partitioned into two sets of (overlapping) subdomains: the power grid partitions that are defined by the black dashed lines; and the boundary windows defined by the blue dashed lines in Figure 5.2. If we discount the fact that an iterative process is actually used to find a "converging" window size for each boundary window, this approach effectively performs multiplicative Schwarz updates for the boundary windows first followed by those for the power grid partitions. If accurate boundary conditions are found in the first step of the approach, the high-level multiplicative Schwarz method essentially converges in one iteration. In this sense, the approach of [126, 128] can be thought as a "smart" multiplicative Schwarz method that exploits the design property of the circuit under analysis.

5.3.3 Parallel Schwarz Preconditioners

Instead of using an iterative Schwarz method to solve the linear system, the same concept can be used to construct a (parallel) Schwarz preconditoner in conjunction with a top-level iterative method such as conjugate gradient and GMRES.

At each top-level iteration, the additive Schwarz preconditioner needs to solve the following linear system to find the solution vector w for a given vector v

$$w = P_{AS}^{-1} v = \sum_{i=1}^{K} (R_i^T A_i^{-1} R_i) v. \tag{5.10}$$

The preconditioner can be parallelized in a way similar to the case of the additive Schwarz method.

The additive overlapping Schwarz preconditoner was used with the GMRES iterative method for analyzing power grids in [104]. Based on an MPI-based parallel implementation on a cluster of four nodes, the additive overlapping Schwarz preconditioning scheme was shown to be more advantageous over the direct solution method and iterative methods employing other preconditioners either in terms of runtime efficiency or memory usage.

5.3.4 Convergence Properties and Parallel Implementation Considerations

While the focus of our discussion is on circuit simulation, it is insightful to understand the convergence properties of the Schwarz methods when applied to linear elliptic PDE problems [103]. The number of iterations to convergence grows with the number of domains. Convergence is poor with zero overlap but improves very quickly with increased overlap.

Finally, the multiplicative Schwarz method requires about half of the iterations of the additive Schwarz method to converge.

A similar dependence of the iterations on the number of subdomains is observed for IC power grid analysis in [128, 126]. For a fixed sized power grid, the size of each subdomain becomes smaller as the number of subdomains increases. If a direct solution method is used for each subdomain, the increase in the number of iterations may be more than compensated by the reduction of runtime spent on factorizing subdomain matrices [126]. This is due to the fact that direct matrix solution has a superlinear time complexity in the problem size.

Several implementation-level considerations exist [103]. While a domain decomposition method has inherent data locality, care needs to be taken in order to have consistent storage and access to the overlapped region. Furthermore, increasing the overlap helps reduce the number of iterations, however, at the cost of increased communication. Similar to the implementation of the parallel Schur method, the circuit needs to partitioned into roughly equally sized subdomains for load balancing. At the same time, partitioning shall be done to minimize the borders between subdomains to reduce the communication overhead.

Finally, while the multiplicative Schwarz method converges faster than the additive counterpart, for efficient parallel implementation of the multiplicative Schwarz method, a coloring scheme with a minimum number of colors is desirable to minimize the number of sequential steps.

6

Parallel Domain Decomposition Methods II

In this section, we discuss domain decomposition methods that are used for solving linearized systems arising from nonlinear circuit analyses and nonlinear domain decomposition methods that operate directly on nonlinear subdomain problems. It shall be noted that from an algorithmic point of view, the distinction between the first class of methods discussed in this section and those in Section 5 may be artificial. This is particularly the case since the methods presented in Section 5 may be immediately applied to facilitate the solution of a nonlinear system by solving the linearized systems arising from the application of the Newton's method. In fact, this approach has been demonstrated in several works.

6.1 Parallel Preconditioning Techniques for Nonlinear Transient Simulation

In [84], the parallel overlapping additive and multiplicative Schwarz methods are used to solve the linearized matrix problem at each Newton iteration for the time domain transient simulation. Large speedups over the serial Berkeley SPICE3 code were demonstrated. The linearized systems resulted from the Newton's method have also been solved using iterative method GMRES as part of the MPI-based parallel simulator Xyce [106]. In this work, the Jacobian matrix was reordered into the block triangular form (BTF) and then the linear system was iteratively solved by using the parallel block-Jacobi (or the nonoverlapping additive Schwarz) preconditioner for GMRES.

In the rest of the section, we will first review related work on parallelizing the harmonic balance method that computes the quasisteady state responses of analog and RF circuits in the frequency domain. Then, we will present a parallel hierarchically preconditioned harmonic balance approach [20, 21, 23, 66]. While this approach can be thought as falling under the category of nonoverlapping Schwarz preconditioning techniques, the frequency domain formulation of the harmonic balance problem does introduce interesting characteristics and complications that are not present in the time-domain circuit analysis. In particular, the Jacobian of harmonic balance has a densely coupled circulant matrix structure. This creates the critical issue of memory-efficient storage of subdomain problems. Furthermore, this approach goes beyond what is discussed earlier in the sense that it is a multilevel hierarchical preconditioning scheme.

Finally, we will briefly review one "intrinsically" nonlinear domain decomposition method, namely, the multilevel Newton method and its application to parallel time-domain circuit simulation [35, 46, 89].

6.2 Hierarchically Preconditioned Parallel Harmonic Balance

Harmonic Balance (HB) is a frequency-domain simulation method for directly computing the steady-state solutions of nonlinear analog and RF circuits with a periodic or quasiperiodic response [63]. The HB formulation leads to large and densely coupled systems of nonlinear equations, for which parallel processing is an attractive solution.

In [95], a method was proposed to partition a circuit into linear and nonlinear portions such that the solution of the linear portion can be parallelized based on the assumption that it dominates the overall runtime. In [93, 94], the work of [95] was extended by exposing the potential parallelism in the form of a directed acyclic graph. Several application-domain-specific methods for allocation and scheduling were discussed. In [56], an implementation of HB analysis on shared memory multicomputers has been reported, where the parallel task allocation and scheduling was applied to device model evaluations, matrix–vector products and the standard block diagonal (BD) preconditioning.

More recently, a hierarchically preconditioned parallel harmonic balance method was reported in [20, 23], where a new multilevel preconditioner was developed and parallelized to improve the robustness and efficiency of the harmonic balance analysis [21, 66]. This approach parallelizes one of the most computationally intensive steps in large-scale HB analysis, preconditioning. Preconditioning is key to HB in that it not only determines the efficiency and robustness of the simulation, but also corresponds to a fairly significant portion of the overall compute work. As a matrix-level preconditioning technique, this approach also admits straightforward integration of traditional parallelizing ideas such as parallelizing device model evaluations, FFT and inverse FFT operations and sparse matrix–vector products.

Following the classifications outlined in Section 2, the algorithmic level of parallel processing is the matrix level and the domain of parallel processing is the frequency domain for the approach in [20, 23]. The latter is the case since the high-level preconditioner is created by partitioning the circuit unknowns along frequency boundaries as detailed in the following subsections.

6.2.1 Problem Formulation of Harmonic Balance

To compute the steady-state response for a nonlinear circuit, the DAE system in Equation (2.1) can be transformed into the frequency domain by using L-point FFT and IFFT as [27, 63]

$$H(X) = \Omega\Gamma q(\cdot)\Gamma^{-1}X + \Gamma f(\cdot)\Gamma^{-1}X - U = 0, \tag{6.1}$$

where X is the Fourier coefficient vector of the circuit unknowns; Ω is a diagonal matrix representing the frequency domain differentiation operator; Γ and Γ^{-1} are the L-point FFT and IFFT matrices, respectively;

$q(\cdot)$ and $f(\cdot)$ are the time-domain charge/flux and resistive equations as defined before; and U is the input excitation expressed in the frequency domain. Assuming that the double-sided FFT/IFFT is used, a total number of $L = 2k + 1$ frequency components are included to represent each signal, where k is the number of positive frequency components being considered.

To solve the above nonlinear system of equations, at each Newton iteration, the Jacobian matrix $J = \partial H / \partial X$ is required

$$J = \Omega \Gamma C \Gamma^{-1} + \Gamma G \Gamma^{-1}, \tag{6.2}$$

where $C = \text{diag}\{c_k = \frac{\partial q}{\partial x}|_{x=x(t_k)}\}$ and $G = \text{diag}\{g_k = \frac{\partial f}{\partial x}|_{x=x(t_k)}\}$ are block diagonal matrices with the diagonal blocks representing the linearizations of $q(\cdot)$ and $f(\cdot)$ at L sampled time points $t_1, t_2, \ldots, t_L$.

Modern harmonic balance algorithms adopt a Krylov-subspace iterative method, such as Generalized Minimal Residual (GMRES) method or its flexible variant (FGMRES) to solve the large linearized problem at each Newton iteration. The use of iterative solvers may not only offer a speed advantage in solving the linearized systems, but also be beneficial in terms of storage. Since the Jacobian matrix is dense, storing it explicitly is very expensive for large circuits. When (F)GMRES is adopted, no explicit storage of the Jacobian is required as long as the matrix–vector products can be provided. The latter can be rather efficiently facilitated by using fast FFT/IFFT and sparse matrix–vector multiplications in the time domain [27].

Preconditioning plays a critical role in determining the efficiency and robustness of HB simulation. The block diagonal preconditioner of [27] works very effectively for mildly nonlinear circuits while losing its efficiency and robustness for highly nonlinear circuits [21, 66]. As will be discussed next, the choice of the preconditioning technique also plays a role in how parallelism may be exploited.

6.2.2 Parallelizable Hierarchical Preconditioning

In [20, 21, 23, 66], a parallelizable hierarchical preconditioner was designed by partitioning the linearized HB problem $JX = B$ defined by the Jacobian, into, say m subproblems, along the frequency boundaries

$$\begin{bmatrix} J_{11} & J_{12} & \cdots & J_{1m} \\ J_{21} & J_{22} & \cdots & J_{2m} \\ \vdots & \vdots & \ddots & \vdots \\ J_{m1} & J_{m2} & \cdots & J_{mm} \end{bmatrix} \begin{bmatrix} X_1 \\ X_2 \\ \vdots \\ X_m \end{bmatrix} = \begin{bmatrix} B_1 \\ B_2 \\ \vdots \\ B_m \end{bmatrix}, \tag{6.3}$$

To more clearly reveal the structure of the Jacobian, J is written in a slightly different form

$$[J]_{m\times m} = \left(\begin{bmatrix} \Omega_1 & & & \\ & \Omega_2 & & \\ & & \ddots & \\ & & & \Omega_m \end{bmatrix} C_c + G_c \right), \tag{6.4}$$

where circulants C_c, G_c are correspondingly partitioned as

$$\begin{aligned} C_c = \Gamma C \Gamma^{-1} &= \begin{bmatrix} C_{c11} & \cdots & C_{c1m} \\ \vdots & \ddots & \vdots \\ C_{cm1} & \cdots & C_{cmm} \end{bmatrix} \\ G_c = \Gamma G \Gamma^{-1} &= \begin{bmatrix} G_{c11} & \cdots & G_{c1m} \\ \vdots & \ddots & \vdots \\ G_{cm1} & \cdots & G_{cmm} \end{bmatrix}. \end{aligned} \tag{6.5}$$

To create a parallelizable preconditioner that is processed on m processors, the off-diagonal blocks of Equation (6.5) are discarded, leading to m decoupled linearized problems of smaller dimensions

$$\begin{cases} J_{11}X_1 = [\Omega_1 C_{c11} + G_{c11}]X_1 = B_1 \\ J_{22}X_2 = [\Omega_2 C_{c22} + G_{c22}]X_2 = B_2 \\ \qquad\qquad \vdots \\ J_{mm}X_m = [\Omega_m C_{cmm} + G_{cmm}]X_m = B_m \end{cases}. \tag{6.6}$$

By solving these decoupled linearized problems in a parallel way, a parallel preconditioner is efficiently established.

This idea of partitioning can be extended in a hierarchical fashion as shown in Figure 6.1. At the topmost level, to solve the top-level linearized HB problem, a preconditioner is created by approximating

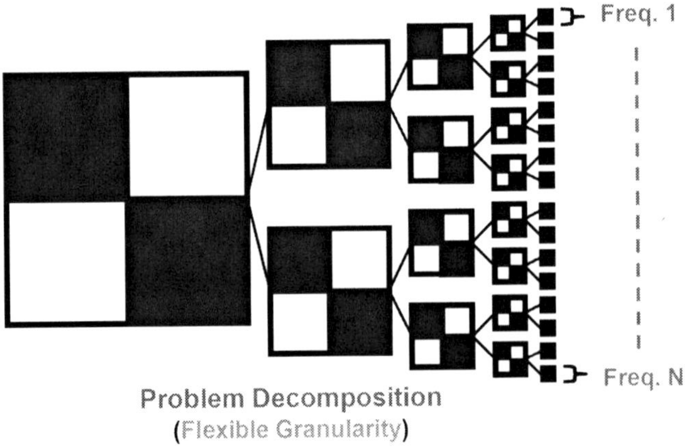

Fig. 6.1 Hierarchical partitioning of the linearized HB problem (from [23] ©[2009] IEEE).

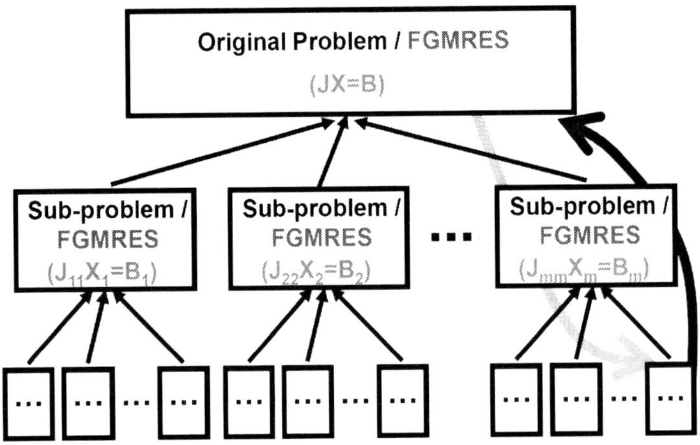

Fig. 6.2 Parallelizable hierarchical HB preconditioner (from [23] ©[2009] IEEE).

the full Jacobian using several (in this case two) *super* diagonal blocks. The size of these superblocks may be large such that to solve them an iterative method such as FGMRES is desirable again. The second-level preconditioners for the iterative solver are created in the same fashion by further partitioning the Jacobian matrix. This entire process essentially leads to a multilevel hierarchical preconditioner for the original linearized HB problem as shown in Figure 6.2.

6.2.3 Implicit Representations of Sub-HB Problems

An important issue in realizing the hierarchical preconditioner is to maintain the implicit nature of matrix storage not only for the top-level Jacobian matrix, but also for all the subproblems. An elegant solution was presented in [21, 66].

As an example, G_{c22}, one matrix block in Equation (6.6), may be approximated by a circulant based on the DFT of some time-domain waveforms defined by G_2, which is a *slower* version of time-varying linearizations G used to define the top-level Jacobian as in Equation (6.2). With this, G_{c22} is approximated by

$$G_{c22} \approx \Gamma_{22} \cdot G_2 \cdot \Gamma_{22}^{-1}, \tag{6.7}$$

where Γ_{22} and Γ_{22}^{-1} represent the DFT and the inverse DFT matrices, respectively with a dimension identical to that of G_{c22}. G_2 is in the time domain and only contains the first few frequency components of G. This implies that by efficiently low-pass filter G to generate G_2 in the time domain, G_{c22} can be implicitly represented by Equation (6.7). This low-pass filtering process is applied several times such that all the subproblems in the hierarchical preconditioner are implicitly defined in the same way [21, 66].

Algorithmically, the hierarchical preconditioner is more advantageous than the standard BD preconditioner [27]. It better approximates the full Jacobian and leads to improved efficiency and robustness, especially for strongly nonlinear circuits. In terms of parallel processing, the hierarchical preconditioner lends itself to natural parallelization. All the subproblems at a particular level are completely independent and hence can be straightforwardly processed in parallel. Furthermore, the structure of the parallel preconditioner, in terms of the depth of hierarchy and sizes (numbers) of subproblems, can be chosen to match a given parallel hardware with a desirable parallelization granularity. This basic idea of hierarchical partitioning was also extended for the harmonic-balance-based analysis of autonomous circuits and envelope-following analysis [23].

The hierarchical preconditioner was shown to be up to fine times more efficient than the BD preconditioner of [27]. The parallel

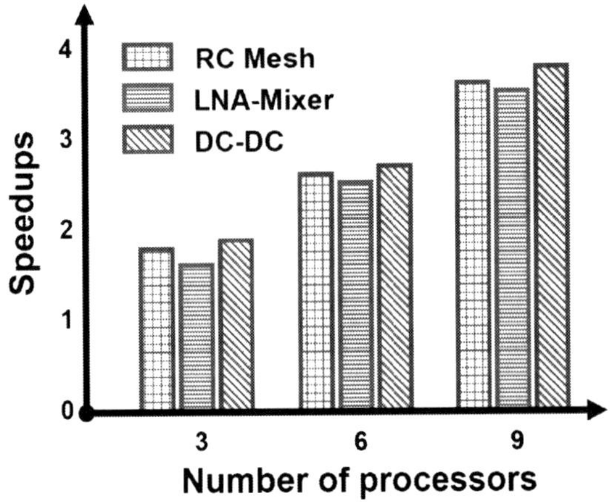

Fig. 6.3 Parallel speedups of the hierarchical HB preconditioner on networked workstations (from [23] ©[2009] IEEE).

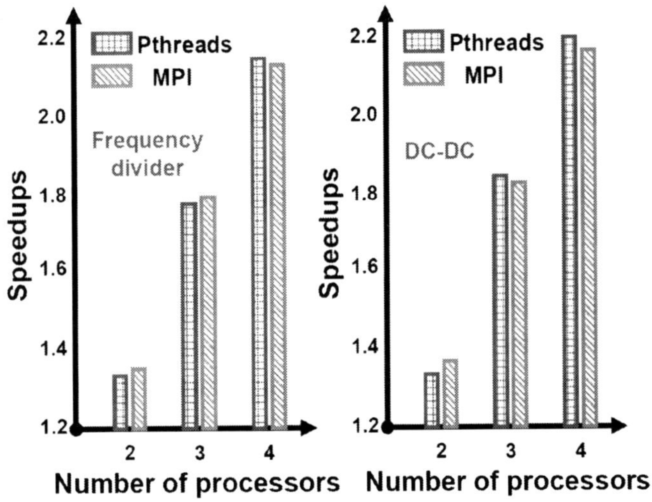

Fig. 6.4 Parallel speedups of the hierarchical HB preconditioner: distributed vs. shared-memory platforms (from [23] ©[2009] IEEE).

implementation further sped up the analysis. Figure 6.3 shows the achieved parallel speedups based on an MPI implementation for three circuits on a network of nine Linux workstations [23].

The achievable parallel speedups are also compared across two hardware platforms: distributed vs. shared-memory in Figure 6.4 [23].

The shared-memory implementation used the pThreads library. It is noted that the speedups obtained from the two platforms were quite similar, due to the coarse-grained nature of the parallel processing exploited in the hierarchical preconditioner.

6.3 Parallel Multilevel Newton Method

The Schur complement approach for solving linear systems presented in Section 5.2 can be extended to the nonlinear domain. This leads to the multilevel Newton method [35, 46, 89]. The algorithmic level of parallel processing of the multilevel Newton method is the nonlinear equation level and it operates in the circuit domain.

As in the case of linear systems, the domain of a circuit is partitioned into $K+1$ nonoverlapping subdomains $D = \{D_1, D_2, \ldots, D_K, \Omega_{\text{int}}\}$, where there is no coupling between subdomains $D_1, D_2, \ldots, D_K$ and each subdomain D_i can couple only with the interface nodes in Ω_{int}. Correspondingly, we have the following partitioned nonlinear systems of equations that arise from the application of a numerical integration to Equation (2.1)

$$f_i(x_i, y) = 0, \quad i = \{1, 2, \ldots, K\} \tag{6.8}$$

$$g(x_1, x_2, \ldots, x_K, y) = 0, \tag{6.9}$$

where the unknown vector x is partitioned into $x = [x_1, x_2, \ldots, x_K, y]^T$ as before. Equation (6.8) corresponds to the local nonlinear equation for subdomain D_i and is processed by a slave compute node. Equation (6.9) is the top-level nonlinear system processed by a single master node.

The parallel Schur complement approach of Section 5.2 can be immediately applied to solve the linearized system at each Newton equation iteration for the nonlinear system in Equation (6.8). However, this simple approach has the following potential limitation. Since several Newton iterations may be needed before the solution to the nonlinear system converges, there may be a considerable amount of interprocessor communication, which limits the efficiency of the parallel simulation.

A tradeoff can be made between communication and computation by introducing the multilevel Newton method. In this case, one Newton iteration consists of an inner iteration loop and an outer Newton update step. In the inner iteration loop, each local nonlinear equation is iteratively solved to convergence under a fixed outer (interface) variable vector y to update all local variables $x_1, \ldots, x_K$. Then an outer Newton step is taken to update outer variable vector y based on the solutions received from the slaves. Finally, to complete one Newton iteration for the entire system, a Newton step is taken to correct all local variables $x_1, \ldots, x_K$ using the updated y.

Since more work is done at the slave level in the above two-level Newton method, the number of top-level Newton iterations may be reduced, leading to less communication between the slaves and the master. The conditions under which the multilevel Newton method maintains local quadratic convergency was provided in [89]. Latency can be also exploited under this multilevel framework [35, 89]. The parallel multilevel Newton method was demonstrated as part of the TITAN simulator, running on both computer clusters and shared-memory multiprocessors [35]. Good parallel speedups were demonstrated up to eight processors.

It is worth noting that since the top-level domain decomposition structure of the multilevel Newton method is identical to the parallel Schur method when applied to solve a linear system as discussed in 5.2, the performance of the multilevel Newton method is also critically dependent on the quality of circuit partitioning. To this end, an ideal partitioning shall lead to a small number of interdomain couplings, hence a small top-level nonlinear equation is defined on the interface domain Ω_{int}. As discussed earlier, this would be the case if the circuit under analysis is strongly structured.

7

Parallel Time-Domain Simulation Using Advanced Numerical Integration Techniques

A variety of time-domain simulation methods have been discussed in the previous sections. Except for waveform relaxation, the main target of all other methods is to solve the nonlinear system of equations arising from the application of a standard numerical integration at each time point, one at a time. In the waveform relaxation method, the exploration of parallelism extends into the functional space of "waveforms", however, with certain limitations in terms of convergence properties.

Parallelism can be also explored along the time axis, namely, by either leveraging numerical integration techniques that are amenable to parallelization, or by creating multiple computation tasks at different time instances that can be processed in parallel. Along the above line, we discuss two recent parallel time-domain simulation techniques [22, 24].

7.1 Waveform Pipelining

To explore parallelism along the time axis for transient circuit simulation, *Waveform Pipelining* (abbreviated as *WavePipe*) was proposed in [24]. *WavePipe* exploits parallelism in the time-domain transient simulation by simultaneously computing circuit solutions at multiple adjacent time points in a way resembling hardware pipelining. Unlike waveform relaxation, without partitioning the circuit, *WavePipe* only exploits parallelism within a few adjacent time points while maintaining stability and accuracy.

The algorithmic level of parallel processing of *WavePipe* is the nonlinear differential equation level. Its domain of parallel processing in the signal domain. There are two pipelining schemes in *WavePipe*: backward (BWP) and forward pipelining (FWP) schemes. BWP creates independent computing tasks that contribute to a larger future time step by moving backward in time while FWP facilitates predictive computing along the forward direction of time. Conceptually, the granularity of parallelization is the solution of the entire circuit at a particular time point. The exploration of this coarse-grained parallelism again can be easily combined with that of other finer grained approaches to achieve greater parallel speedups.

7.1.1 Backward Waveform Pipelining (BWP)

The main challenge in parallelizing transient analysis along the time axis is the data dependency in time. As depicted in Figure 7.1, standard transient analysis solves the circuit response sequentially in the forward direction of time, where the known circuit responses at preceding points provide an initial condition for computing the response further down the time axis. Conceptually, BWP "breaks" this data dependency via a new way of looking at multistep numerical integration methods as follows.

The multistep Gear's integration formula is given by [38]

$$x_{n+1} = \beta_0 \dot{x}_{n+1} + \sum_{k=1}^{p} \alpha_k x_{n+1-k}, \tag{7.1}$$

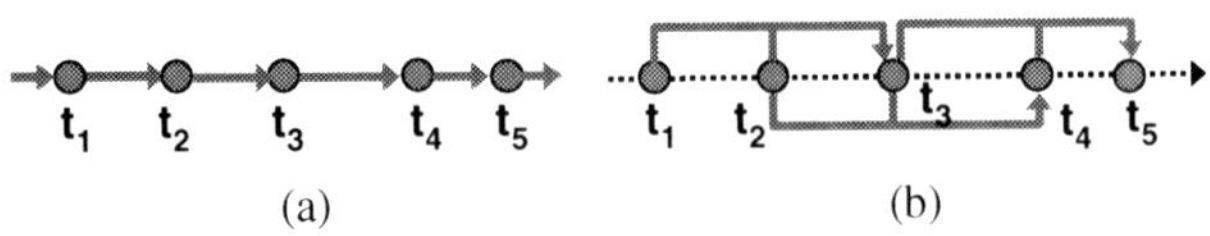

Fig. 7.1 Data dependency in (a) one-step and (b) multistep (two-step) numerical integration (from [24] ©[2009] IEEE).

where p is the order of numerical integration, $x_i, i = n+1-p, \ldots, n$ is the circuit response at time point i, x_{n+1} is the unknown circuit response at time point $(n+1)$, and (β_0, α_k) are certain coefficients, which are constant in fixed-step Gear's methods. Consider the special case of the two-step variable time-step Gear's method [101]

$$\begin{aligned} x_{n+1} &= -x_{n-1}\frac{h_{n+1}^2}{h_n(2h_{n+1}+h_n)} + x_n\frac{(h_{n+1}+h_n)^2}{h_n(2h_{n+1}+h_n)} \\ &\quad + \dot{x}_{n+1}\frac{h_{n+1}(h_{n+1}+h_n)}{2h_{n+1}+h_n}, \end{aligned} \tag{7.2}$$

where $h_{n+1} = t_{n+1} - t_n$, $h_n = t_n - t_{n-1}$. The local truncation error of Equation (7.2) is

$$\varepsilon_{n+1} = -\frac{h_{n+1}^2(h_{n+1}+h_n)^2}{6\cdot(2h_{n+1}+h_n)} \cdot x^{(3)}(\tau), \tag{7.3}$$

where τ is in $[t_n, t_{n+1}]$, and the third order derivative $x^{(3)}(\tau)$ may be approximated as $x^{(3)} \approx 3!DD_3(t_{n+1}, t_n, t_{n-1}, t_{n-2})$, in which $DD_3(t_{n+1}, t_n, t_{n-1}, t_{n-2})$ denotes the third order divided difference evaluated at time points $t_{n+1}, t_n, t_{n-1}, t_{n-2}$.

The commonly adopted local truncation error (LTE)-based time-step control schemes selects the largest stepsize that does not exceed a specified local truncation error tolerance. This strategy ensures a good control of amount of local error incurred at each time step while pushing the transient simulation forward as fast as possible. In the case of two-step Gear's (Gear2) method, the stepsize can be estimated based on LTE as follows. For a given $h_n = t_n - t_{n-1}$, $DD3$ and a specified LTE tolerance ε, denote the stepsize to be determined by $h_{n+1} = kh_n, k > 0$. According to Equation (7.3), the maximum allowable h_{n+1} is determined by the following equation

$$\frac{k^2(k+1)^2}{(2k+1)} = \left|\frac{\varepsilon}{DD_3\cdot h_n^3}\right|. \tag{7.4}$$

7.1.1.1 Parallel Backward Waveform Pipelining

To see how parallelism can be explored, consider the case of running two concurrent threads and adopting the two-step variable-stepsize Gear's integration method. To simplify the discussion, for the purpose of

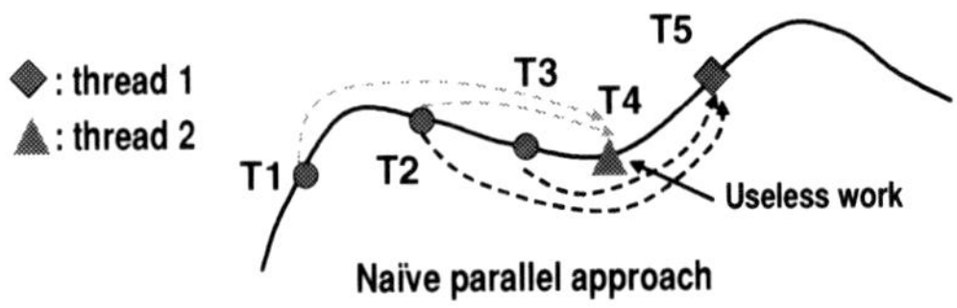

(a)

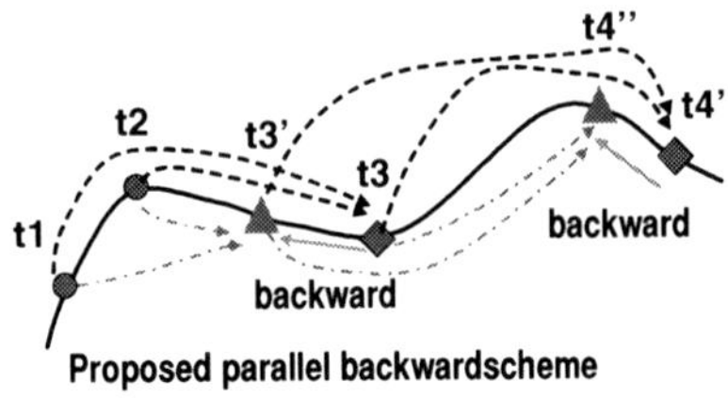

(b)

Fig. 7.2 Parallel double-threaded backward pipelining: (a) a naïve approach and (b) more effective backward pipelining (from [24] ©[2009] IEEE).

estimating the LTE, the third order divided difference $DD3$ is assumed to be constant at different time points. In this case, the LTE is only a function of the two time steps, h_n and h_{n+1}, in Equation (7.3).

A naïve parallel approach is shown in Figure 7.2 (a). The circuit responses at three time points T_1, T_2, and T_3 are assumed to be known. Using the solutions at T_2 and T_3 as the initial condition, one may use the first thread to compute the solution at T_5. Concurrently, the second thread may be launched to compute the solution at T_4 by using solutions at T_1 and T_2 as the initial condition. While seemingly some concurrence is created, however, such concurrence would be of little use in practice. Unless in rare cases, the work done by the first thread at T_5 would be usually beyond that of the second thread at T_4 as the first thread uses the most updated initial condition.

To create more useful parallel work, let us examine the dependency of the LTE of the two-step numerical integration formula, as given in Equation (7.3), on h_n and h_{n+1}, respectively. The partial derivatives of the LTE with respect to h_n and h_{n+1} are given by

$$\begin{aligned} \frac{\partial|\varepsilon|}{\partial h_{n+1}} &= \frac{2|DD_3|[h_{n+1}(h_{n+1}+h_n)\cdot(h_{n+1}^2+(h_{n+1}+h_n)(2h_{n+1}+h_n))]}{(2h_{n+1}+h_n)^2} \\ \frac{\partial|\varepsilon|}{\partial h_n} &= \frac{|DD_3|h_{n+1}^2(h_{n+1}+h_n)(3h_{n+1}+h_n)}{(2h_{n+1}+h_n)^2} \end{aligned} \tag{7.5}$$

Both derivatives are positive, which implies that the LTE increases with both h_n and h_{n+1} as expected.

This observation suggests the following approach to leverage parallelism, as depicted in Figure 7.2 (b). Given the known initial conditions at t_1 and t_2, the first thread is assigned to compute the circuit solution at t_3, determined by the LTE-based stepsize control. In parallel, the second thread is started to work on the circuit response at t'_3, which is located between t_2 and t_3, using the same initial condition. We refer to the computation of the circuit response at t'_3 as a backward step relative to t_3. It shall be noted that if the LTE tolerance is satisfied at t_3, practically, the same LTE tolerance will be satisfied at any placement of t'_3 that is between t_2 and t_3.

Now the question is that why the work done at t'_3 would be useful. This can be understood as follows. Upon the completion of both threads, the solutions computed at t_3 and t'_3 could be used as the initial condition to compute the circuit response at the next time point in the future. Compared with the serial transient simulation where the solutions at t_2 and t_3 are provided as the initial condition, the known circuit response at t'_3 helps to reduce the value of h_n in the LTE-based step control as defined in Equation (7.4). Hence, it is likely that the use of this backward step may lead to a larger stepsize in the future and advance the transient analysis faster along the time axis. The same way of interleaving the work of these two threads is repeated for future time points such as t_4 (t'_4) in Figure 7.2(b). In practice, the placement of t'_3 (or t'_4) can be tuned experimentally to optimize the runtime performance [24].

In principle, the double-threaded backward pipelining scheme can be generalized for multistep methods. For example, a three-thread parallel scheme where two threads are launched to simultaneously compute the circuit solutions at two backward steps can further improve future stepsizes. However, such additional performance improvement is usually limited in practice.

7.1.2 Forward Waveform Pipelining (FWP)

Parallelism can be also explored in the forward direction of the time in FWP. To see this, consider the situation shown in Figure 7.3. Again,

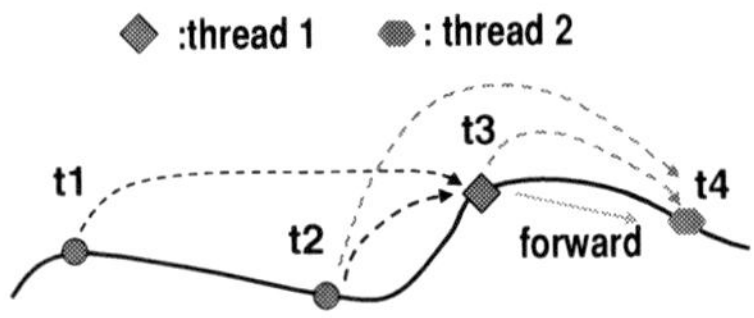

Fig. 7.3 Double-threaded forward waveform pipelining (from [24] ©[2009] IEEE).

the circuit responses at time points t_1 and t_2 are assumed to be known and treated as a given initial condition. A seemingly useful approach is to use the LTE-based stepsize to determine a suitable stepsize for a two-step integration method (e.g., the Gear2 method) such that the first thread may be allocated to compute the circuit response at t_3 using the solutions at t_1 and t_2 as the initial condition. Then, the second thread is assigned to compute the response at a time point t_4 that is further down the time axis using the same initial condition. However, the problem is that the solution obtained at t_4 may not be trusted if the maximum permissible stepsize is already employed at t_3.

The above problem is remedied in the forward scheme as shown in Figure 7.3. While the first thread solves the circuit response at t_3 using the known solutions at t_1 and t_2 as the initial condition, useful parallelism would be explored if the second thread were started to find the response at t_4 using the responses at t_2 and t_3 as the initial condition in parallel. Obviously, this creates yet another problem due to the data dependency between the two threads since the circuit response at t_3 may not be available yet as it is computed by the first thread.

The data dependency described above is resolved as follows. To start the work at t_4 concurrently, not only does the solution at t_3 need to be known, but also the stepsize $h_f = t_4 - t_3$ must be decided. However, h_f also depends on the unknown circuit solution at t_3, say $x(t_3)$. This difficulty is resolved by efficiently estimating $x(t_3)$ and using such estimate to launch the second thread. In particular, Forward Euler (FE) or linear extrapolation may be used to compute a quick estimate, say $\tilde{x}(t_3)$.

Since the error may be incurred in such estimate, several complications arise and shall be addressed. Using $\tilde{x}(t_3)$ may lead to an overly optimistic stepsize h_f, which may render the work done at t_4 violating the LTE tolerance, and hence useless. To reduce the chance of stepsize overestimation, a damping factor β ($\beta < 1.0$) is introduced to scale down the estimated stepsize such that a more conservative stepsize is used: $h_{f,damped} = \beta h_{f,FE}$. Once the exact response at t_3 becomes available, the LTE is recomputed. If the LTE exceeds the tolerance, the work done at t_4 will have to be revoked [24].

More importantly, the circuit solution computed in parallel at t_4 using $\tilde{x}(t_3)$ may not be accurate even if the time step is estimated conservatively. In [24], two interthread communication schemes were proposed to guarantee the accuracy of the solution, as shown in Figure 7.4. Figure 7.4(a) depicts a coarse-grained approach in which at least one nonlinear Newton iteration is performed at t_4 after the solution at t_3 has fully converged. This guarantees that the transient response at t_4 computed by thread 2 will converge to the exact value using the accurate response at t_3. The synchronization cost is one interthread communication by which thread 2 fetches the converged $x(t_3)$ from thread 1.

In the finer grained approach of Figure 7.4(b), thread 2 more frequently updates $\tilde{x}(t_3)$ as thread 1 works toward the convergence. The most recent update of $\tilde{x}(t_3)$ that is available at the end of each Newton

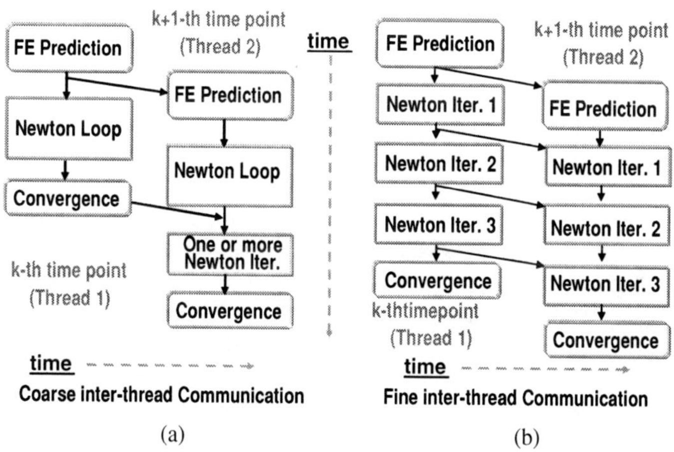

Fig. 7.4 Fine vs. coarse-grained interthread communications to guarantee the accuracy of the parallel forward waveform pipelining (from [24] ©[2009] IEEE).

iteration in thread 1 may be loaded to thread 2 to accelerate its convergence. As before, to ensure the final simulation accuracy, once $x(t_3)$ fully converges, it will be loaded by thread 2 as part of the exact initial condition to perform one or more Newton iterations to finally converge on the solution at t_4. Essentially, communication overhead is traded off for the number of Newton iterations needed for convergence. This tradeoff is advantageous for large circuits where one Newton iteration may be much more expensive than an interthread communication.

7.1.3 Combined BWP/FWP Schemes and Thread Scheduling

The backward and forward schemes can be combined to create a variety of multithreaded WavePipe implementations to utilize a larger number of processor cores. WavePipe can be further combined with low-level parallel schemes such as parallel device model evaluations and matrix solvers for a higher level of parallelism.

As an example, a multithreaded WavePipe configuration utilizing four threads is shown in Figure 7.5. This configuration contains one base thread, one backward (BWP) thread, and two forward (FWP) threads. Starting from the two known circuit responses (used as the initial condition), base thread T1 starts to work on the transient response at the next time point, referred to as the base point, using a standard two-step integration method such as Gear2. Before actually solving the response at the base point, T1 first determines the stepsize and then computes a quick estimate for the unknown response (e.g., via extrapolation). This

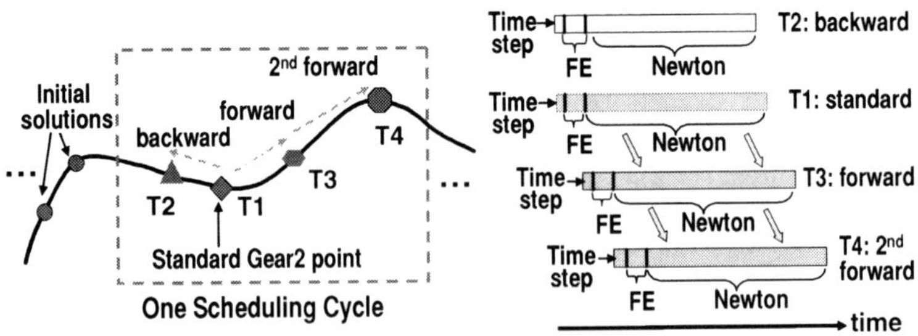

Fig. 7.5 Four-thread waveform pipelining (from [24] ©[2009] IEEE).

estimate is employed to enable the forward pipelining scheme. As T1 continues to execute, the second thread T2 is launched to simultaneously solve for the transient response at a different time point (referred to as the backward point) according to the backward pipelining scheme. In the meantime, using the estimated solution computed by T1, the third thread T3 is started to solve the circuit response at a point further down the time axis (referred to as a forward point) according to the FWP scheme. In a similar fashion, T3 may also first estimate the unknown transient response. This estimate is used to launch the second FWP thread, T4.

The launching and the completion of these four threads is referred to as a *thread scheduling cycle*. It is useful to note that within one scheduling cycle, T2 may complete slightly before T1. This is the case since a smaller stepsize is used for the work of T2. The data dependency between T1, T3, and T4 forces them to complete subsequently.

Upon the completion of a scheduling cycle, a new cycle starts in the same fashion. In the case of Figure 7.6(a), where all the threads successfully complete in the cycle, the solutions computed by T3 and T4 will be used as the initial condition to start the next cycle. For the FWP scheme, any stepsize overestimation will lead to discarding the computed work. In Figure 7.6(b), if T3 overestimates its stepsize, the work done by T3 will not satisfy the LTE tolerance and shall be discarded. Due to the data dependency between T3 and T4, the work done by T4 shall be discarded as well. In this case, the next scheduling cycle will

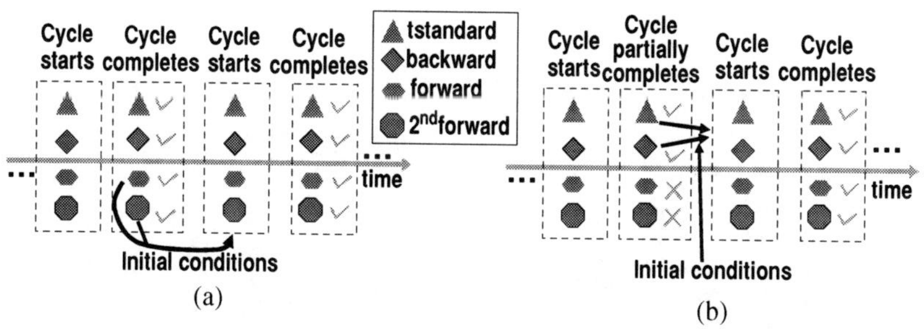

Fig. 7.6 4T waveform pipelining: (a) without revoking of forward pipelining and (b) with revoking of forward pipelining (from [24] ©[2009] IEEE).

be started by using the solutions computed by T1 and T2 as the initial condition.

In practice, modifications may be made to the thread scheduling policies outlined above. The optimal policy together with the optimal values of various control parameters may be experimentally selected.

For several test circuits, the parallel speedups of six WavePipe schemes are shown in Figure 7.7 as reported in [24]. The hardware used was a shared-memory workstation consisting of two Intel Xeon quadcore processors. These six schemes examined were: two-thread BWP, two-thread FWP, three-thread 1-BWP/1-FWP, three-thread two-FWP, four-thread 1-BWP/2-FWP, and four-thread 3-FWP schemes, respectively. Each of these scheme further included a base thread. The speedups were defined w.r.t the runtimes of the standard serial SPICE-like circuit simulation using backward Euler method. The average runtime speedups of the six schemes were shown to be 1.33x, 1.62x, 1.80x, 1.89x, 2.14x, and 2.26x, respectively.

The proposed coarse-grained parallel WavePipe was compared with the low-level scheme that uses parallel transistor device model evaluation and matrix solution (Figure 7.8). The public domain parallel matrix solver SuperLU [16] was employed. The comparison was made using a double-balanced mixer and an RLC mesh circuit. When the number of threads was in between two and four, only the high-level waveform pipelining was employed in WavePipe. The eight-thread implementation, on other hand, combined the three-forward waveform pipelining and parallel device model evaluation and matrix solution. In particular, within each mode of operation (one base mode, three forward modes), two threads were utilized to facilitate device model

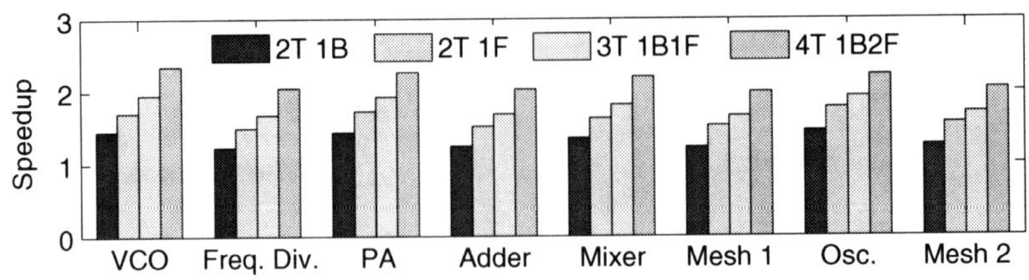

Fig. 7.7 Parallel speedups of several WavePipe schemes (from [24] ©[2009] IEEE).

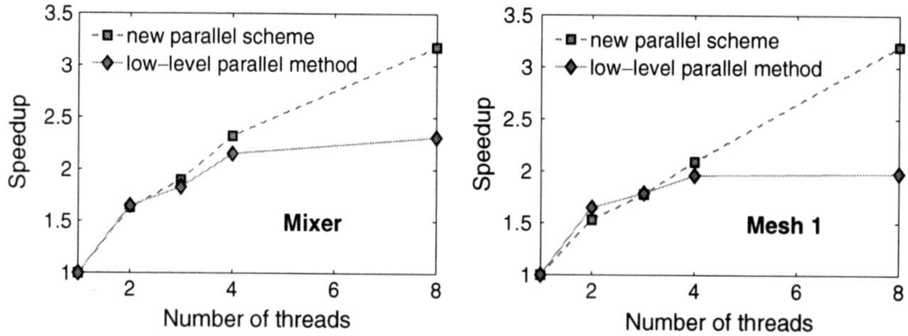

(a) Double-balanced mixer

(b) RLC mesh

Fig. 7.8 Hybrid WavePipe vs. traditional parallel model evaluation/matrix solving (from [24] ©[2009] IEEE).

evaluation and matrix solving. When the number of threads varies from two to four, WavePipe was comparable to the low-level parallel scheme. However, the runtime scaling of the low-level scheme started to saturate beyond four threads. The eight-thread parallel model evaluation/matrix solving scheme did not further improve the runtime. This clearly underscores the need to develop orthogonal parallelizing avenues such as WavePipe so as to well utilize the parallel hardware.

7.2 Parallel Telescopic Projective Numerical Integration

The second family of parallel transient simulation approaches we present in this section is based upon the use of advanced explicit numerical integration. It is in contrast with other approaches that have been reviewed, which adopt a (stable) implicit numerical integration method (e.g., backward Euler). From the point view of parallel circuit simulation, the use of explicit numerical integration offers two parallel opportunities. Explicit numerical integration may be leveraged to effectively decompose the circuit into smaller pieces such that the transient response at each time point may be obtained by solving multiple smaller nonlinear equations in parallel. This corresponds to performing parallel processing at the level of the nonlinear equation in the domain of circuits.

While the benefits are clearly visible, the implementation of these ideas is contingent upon the extent to which the stability limitation of classical explicit numerical integration methods may be addressed as described in the following subsections.

7.2.1 Exploration of Stable Explicit Numerical Integration for Parallel Simulation

Explicit numerical integration methods have been attempted for circuit simulation but with limited adoption due to the stability limitation. For example, the explicit integration-based technique called ACES developed in [19] adopts a number of means to alleviate the stability limitation of explicit integration. The heuristic nature of the approach, however, confines its application only to digital timing simulation. The A-stable feature of the high order explicit numerical integration method suggested in [42] is also undermined by the difficulty in applying it to practical circuit simulation problems. However, if successful, the application of explicit methods may offer an efficiency advantage by relaxing the need for solving coupled systems of equations.

From the point view of parallel processing, in the forward direction of time, stable explicit numerical integration methods would enable the parallelization of the standard transient circuit analysis; the application of explicit numerical integration leads to decoupled node-based or block-based systems of equations that may be processed in parallel. Interestingly, the recent development of projective integration methods in the numerical analysis community has provided a basis for addressing some of the stability limitations of explicit methods [39, 40]. Ideas for integrating backward in time have also appeared [41], although a greater care must be taken as maintaining stability becomes more challenging in this case. Approaches leveraging these developments for parallel circuit simulation have emerged as well [22].

7.2.2 Telescopic Projective Integration

Widely distributed time constants or eigenvalues make a circuit stiff. For practical simulation purposes, it is often the case that only the long term transient behavior of the circuit is of interest, which is primarily determined by the slow transients (corresponding to large time constants). Since it is sufficient to only track the slow components in the transient response, it is desirable to select a stepsize that is comparable to large time constants of the circuit to gain simulation efficiency. However, the stiffness of the system limits the use of a standard explicit integration method with a small stability region; the stepsize must be made small (comparable to the smallest time constant of the circuit) to guarantee the stability. While per step cost of explicit numerical integration may be low, the need to track the long-term transient behavior with tiny stepsizes can make the simulation prohibitively expensive. This is particularly true when the ratio between the largest and smallest time constants is large. This is also the reason that an implicit integration method (such as one that is A-stable) is preferred for stiff simulation problems.

The projective integration method and its telescopic extension were developed to alleviate the stability limitation of classical explicit numerical integration methods [39, 40]. To see how they work, consider a stiff circuit whose eigenvalue distribution is shown in Figure 7.9. Note that the eigenvalues, corresponding to the inverses of the time constants, spread out and there is a gap between two clusters of the eigenvalues, corresponding to the fast and the slow components in the system.

The projective integration method works in the following way to ensure the stability of explicit numerical integration [39], as shown in Figure 7.10. In each effective projective integration step, first, an inner integrator (e.g., one that is based on forward Euler) is used to integrate for $k + 1$ steps from time t_n to time t_{n+k+1} to obtain the solutions from x_n to x_{n+k+1}. The stepsize of the inner integrator is made small and a

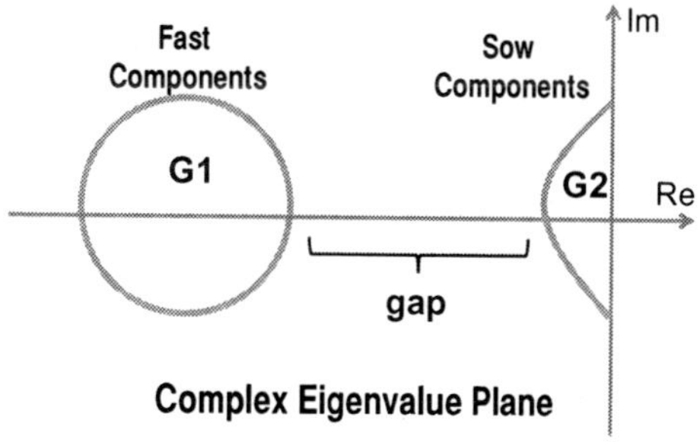

Fig. 7.9 Eigenvalue distribution with a gap.

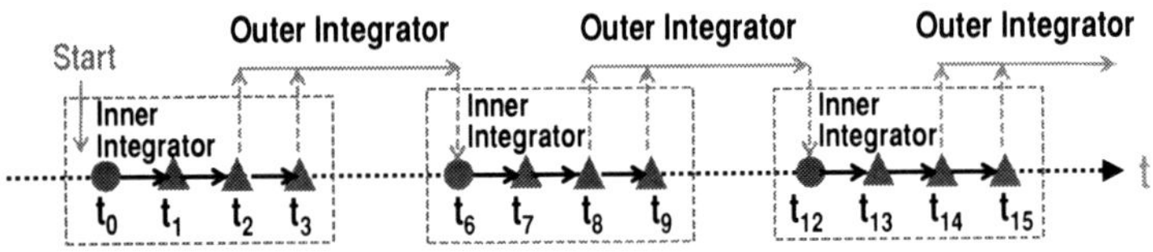

Fig. 7.10 Projective numerical integration forward in time.

sufficient number of steps are taken such that fast transient responses die out in the stiff system. As a result, the accumulated error introduced is sufficiently damped and there is no concern on the stability issue caused by the fast components of the system. Then, a projective step, which forms the outer integrator, extrapolates forward over M inner steps to get $x_{n+k+1+M}$ using solutions x_{n+k} and x_{n+k+1}

$$x_{n+k+1+M} = (M+1)x_{n+k+1} - Mx_{n+k}. \tag{7.6}$$

The large extrapolation step has a stepsize commensurate with the slow time constants, hence, boosting the efficiency of the overall integration scheme. This step may not cause any stability problem if the inner integrator has sufficiently damped the error, as analyzed later. The combination of several small integration steps (the inner integrator) and one large extrapolation step (the outer integrator) can be deemed an effective step of the projective integration. Note that an explicit method such as forward Euler can be employed for the inner integrator. As long as the stepsize of the inner integrator is well controlled, the accumulated error will be sufficiently damped. The large extrapolation step at the very end is also explicit, making the entire projective integration scheme explicit in nature. On the other hand, due to the stepsize boosting by the outer integrator, the effective (or average) stepsize of a projective integration step is no longer limited by that of the inner integrator and can be made comparable to the time scales of slow components in the circuit.

The projective integration scheme is outlined in Algorithm 3, where the inner integrator and the outer integrator are denoted by $I_{inner}(\cdot)$ and $I_{outer}(\cdot)$, respectively. The inner integration from solution x_i at time t_i to solution x_{i+1} at time t_{i+1} is denoted by $x_{i+1} = I_{inner}(x_i)$. The output of one projective step, starting from solution y_j, is denoted by y_{j+1} (j is the index at the outer integration level) and represented by $y_{j+1} = I_{outer}(y_j)$.

Algorithm 3: Forward-Time Projective Integration

inputs : y_j, k, M and $I_{inner}(\cdot)$
output: $y_{j+1} = I_{outer}(y_j)$

```
let x_0 = y_j;
for i ← 1 to k + 1 do
  | x_i = I_inner(x_{i-1});
end
y_{j+1} = (M + 1)x_{k+1} − M x_k;
```

The projective integration described above is most suitable for circuits whose eigenvalue distribution has one gap between two clearly separated clusters, as shown in Figure 7.9. In reality, the eigenvalues may be widely distributed without any clear clustering. It is also difficult to know the distribution of the eigenvalues *a priori*. Under this situation, the stepsize of the outer projective step must be significantly constrained ($M < 3k$) to ensure the [0, 1] stability [39], which deteriorates the efficiency of projective integration. To remedy this problem, telescopic projective integration approach was developed [40]. The key idea is to recursively apply the concept of projective integration under a multilevel integration framework. At each level, to ensure the stability, the outer projective step size is constrained as before, leading to a limited stepsize amplification of $(M + k + 1)/(k + 1)$ with a constrained M. However, although a limited speedup is obtained at each level, a significant overall stepsize amplification of $(M + k + 1)^q/(k + 1)^q$ can still be achieved in a q-level telescopic framework. The two-level telescopic projection scheme is illustrated in Figure 7.11.

Using the same notations of Algorithm 3, the telescopic projective integration is described in Algorithm 4, where I_0 is the bottom-level inner integrator and $I_i, (i > 0)$ is the i-th level outer integrator. y_j^q and y_{j+1}^q are the circuit responses at the two adjacent time points on the q-th level of the integration scheme.

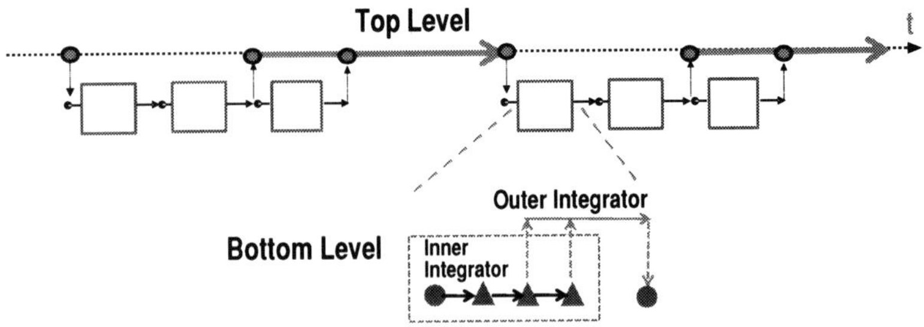

Fig. 7.11 Two-level telescopic projective numerical integration forward in time.

Algorithm 4: q-level Forward-Time Telescopic Projective Integration

inputs : y_j^q, k, M and $I_{q-1}(\cdot)$
output: $y_{j+1}^q = I_q(y_j^q)$

let $z_0 = y_j^q$;
for $i \leftarrow 1$ **to** $k+1$ **do**
  $z_i = I_{q-1}(z_{i-1})$;
end
$y_{j+1}^q = (M+1)z_{k+1} - Mz_k$;

7.2.3 Stability Analysis of the Telescopic Projective Integration-Based Circuit Simulation

To shed more light on the stability property of telescopic projective integration when applied to standard transient circuit simulation [22], consider the following linear homogenous system as a test problem

$$Gx(t) + C\frac{dx(t)}{dt} = 0, \tag{7.7}$$

where $x(t) \in \mathbb{R}^N$ and $G, C \in \mathbb{R}^N$. For ease of discussion and without loss of generality, it is assumed that $G^{-1}C$ is diagonalizable as $P\Lambda^{-1}P^{-1}$, where Λ is diagonal. We further assume C^{-1} exists, which practically implies that there is a grounded capacitance incident to each node.

The inner integrator for projective integration is considered to be forward Euler. Integrating for $k+1$ inner integration steps in the forward direction of time gives the solutions after k and $k+1$ inner steps as

$$\begin{cases} X_{n+k} = P(I - h\Lambda)^k P^{-1} \cdot X_n \\ X_{n+k+1} = P(I - h\Lambda)^{k+1} P^{-1} \cdot X_n \end{cases}, \tag{7.8}$$

where h is the stepsize of the inner integrator. The extrapolated solution at $t_{n+k+1+M}$ is determined by combining Equations (7.8) and (7.6)

$$\begin{aligned} X_{n+k+1+M} &= P((M+1)(I - h\Lambda)^{k+1} - M(I - h\Lambda)^k)P^{-1}X_n \\ &= P\Phi_f P^{-1} X_n, \end{aligned} \tag{7.9}$$

where $\Phi_f = (M+1)(I - h\Lambda)^{k+1} - M(I - h\Lambda)^k$.

Without loss of generality, a two-level telescopic projective integration framework is considered. For the forward telescopic projective integration, the solution at time point $t_{n+(k+1+M)^2}$, which is at the end of telescopic projective integration at the second level, can be represented as

$$X_{n+(k+1+M)^2} = (M+1)X_{n+(k+1)(k+1+M)} - MX_{n+k(k+1+M)}. \tag{7.10}$$

Perform a coordinate transformation and define new state variables Y using P as: $Y = P^{-1}X$, then it is not difficult to show

$$Y_{n+(k+1+M)^2} = ((M+1)\Phi_f^{k+1} - M\Phi_f^k)Y_n. \tag{7.11}$$

Note that Φ_f is a diagonal matrix. Denote its i-th diagonal entry by $\phi_{f,i}$, the corresponding components of $Y_{n+(k+1+M)^2}$ and Y_n by $y_{n+(k+1+M)^2,i}$ and $y_{n,i}$, respectively. Equation (7.11) gives the following

$$\phi_{f\_tel,i} = \frac{y_{n+(k+1+M)^2,i}}{y_{n,i}} = (M+1)\phi_{f,i}^{k+1} - M\phi_{f,i}^k, \tag{7.12}$$

where $\phi_{f,i} = (M+1)(1 - h\lambda_i)^{k+1} - M(1 - h\lambda_i)^k$, and λ_i is the i-th diagonal entry of Λ, correspondingly.

The stability region of the forward telescopic projective integrations is defined by $|\phi_{f\_tel}| \leq 1$. It can be shown that with suitable choices of k and M both stability conditions can be satisfied regardless of the eigenvalue distribution (λ_i's). A procedure similar to what is adopted in [40] can be used to formally show that there exists a $[0,1]$ stability region. This result implies that for any real $\lambda_i \in [0, \frac{1}{h}]$, the telescopic projective method can maintain stability with certain choices of parameters k and M.

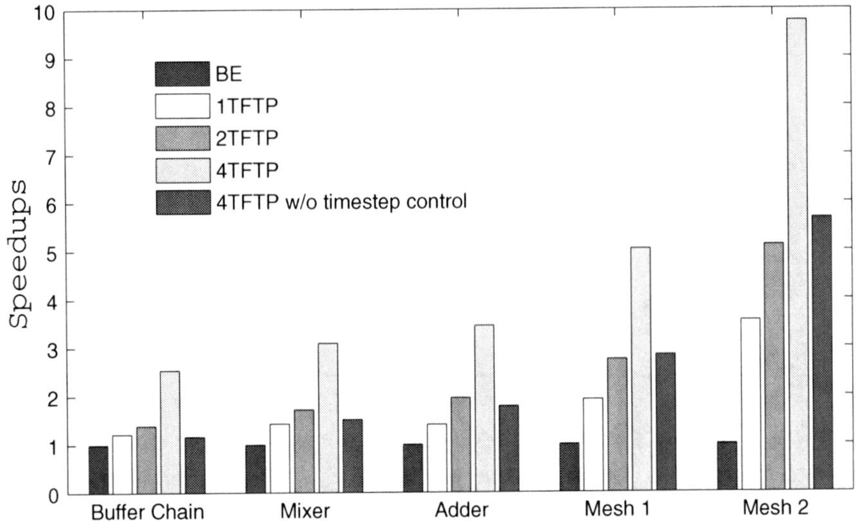

Fig. 7.12 Speedups of the forward-time telescopic projective integration schemes.

7.2.4 Results of Parallel Telescopic Numerical Integration

The efficiency of several multithreaded implementations of the forward-time telescopic projective integration method is examined for five circuits: a buffer chain, a double-balanced mixer, an adder, and two RC mesh circuits, respectively. The hardware platform is a shared-memory Linux server with two quad-core Intel processors. The runtime speedups are defined with respect to the standard serial transient simulation with backward Euler (BE) integration.

The implementations considered include serial (forward-time) telescopic projective integration (1T-FTP), and its two-thread (2T-FTP) and four-thread (4T-FTP) implementations. The number of telescopic level is two and a dynamic time stepping scheme is also employed for the above FTP implementations. For comparison, the dynamic stepping scheme has also been turned off for the four-thread FTP implementation. The runtime results are shown in Figure 7.12, which demonstrates the good speedups obtained by various FTP implementations.

8

Time-Domain Multialgorithm Parallel Circuit Simulation

Current industry trends clearly point to a continuing increase in the number of processors or cores, hence the amount of parallel compute power, in modern computing systems. Thus, it is desirable to create a rich set of parallelization opportunities algorithmically so as to fully utilize the available hardware parallelism. One key observation is that most of the existing parallel circuit simulation approaches, as surveyed in the previous sections, can be viewed as exploring *intraalgorithm parallelism*, where parallel computing is only applied to expedite intermediate computational steps within a single algorithm. While clearly meaningful in its own right, pursing intraalgorithm parallelism often leads to fine-grained parallel processing and may require significant effort in data dependency analysis and parallel programming implementation.

Recently, a time-domain coarser grained multialgorithm parallel circuit simulation (MAPS) framework [119, 120] and its hierarchical extension (HMAPS) [121] have been demonstrated. Unlike conventional approaches where a single (parallel) algorithm is employed for a given application, in MAPS, multiple algorithms with varying characteristics are launched to process the same simulation task. Parallel speedups are obtained by having these algorithms interact with each other in a cooperative manner on the fly. This opens up a somewhat unorthodox angle to approach parallel circuit simulation as it allows one to explore a combination of intra- and interalgorithm parallelism. This combination of different levels of parallelism not only opens up new opportunities, but also allows us to explore advantages that are simply not possible when working within one fixed algorithm. Serial circuit simulation, single-algorithm and multialgorithm parallel circuit simulation are contrasted in Figure 8.1.

Circuit behaviors and activities to be simulated are complex functions of circuit types, structures, and input excitations. In addition, for a fixed circuit, the circuit behavior may vary significantly over the time, exhibiting varying amount of switching activities and exercised nonlinearities. It is not difficult to foresee that the characteristics of circuit behavior have a definite influence on how such characteristics may be simulated by different families of simulation algorithms. For instance, latent phases of the circuit operation may be more efficiently simulated by a fixed Jacobian-type iterative method such as the successive chord method [15]. However, such nonrobust method may lead to slow convergence or convergence failure for highly active phases of a

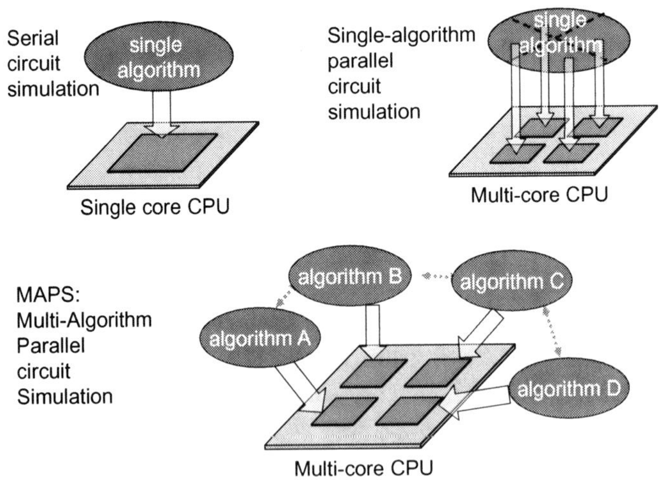

Fig. 8.1 Explore interalgorithm parallelism via multialgorithm simulation.

nonlinear transistor circuit. However, in practice, it is difficult to select a single best algorithm that fits all circuit types or even one complete simulation run for a given circuit.

This observation of the performance variation of a single simulation algorithm over the time and circuit types suggests potential benefits that may be gained from running multiple algorithms in parallel. Along this line, MAPS leverages multiple algorithms, ideally with varying tradeoffs between runtime efficiency and robustness, to explore algorithm diversity. A standard SPICE-like simulation algorithm, which is robust but potentially slow, is always chosen in the algorithm pool to guarantee the success of the simulation. Hence, the worst case performance of MAPS is typically not worse than a serial SPICE simulation.

As discussed earlier, a nonlinear circuit can be described by the DAE in Equation (2.1). It is evident from Figure 2.1 that numerical integration and iterative solution of nonlinear equations are the two critical components of time-domain transient analysis. In MAPS, different numerical integration methods are paired with different nonlinear iterative methods to create a candidate algorithm pool.

8.1 Algorithmic Diversity in Numerical Integration

Several one-step and multistep integration methods are adopted in MAPS to explore varying tradeoffs between the computational cost and allowable step size.

8.1.1 One-Step Integration Methods

Backward Euler and Trapezoidal are one-step integration methods in that they rely on the circuit solution at one preceding time point. The two integration formulas are defined as $x_{n+1} = x_n + h_{n+1}x'_{n+1}$ and $x_{n+1} = x_n + \frac{h_{n+1}}{2}(x'_n + x'_{n+1})$, respectively, where x_{n+1} and x_n are the values of a circuit variable at $(n+1)$-th and n-th time points, and h_{n+1} is the stepsize for the $(n+1)$-th step, respectively.

The local truncation errors (LTEs) at time t_{n+1} introduced by BE and TR are given as

$$LTE_{BE} = -h_{n+1}^2 \frac{x''(\xi)}{2}, \quad LTE_{TR} = -h_{n+1}^3 \frac{x'''(\xi)}{12}, \tag{8.1}$$

where $t_n \leq \xi \leq t_{n+1}$. Variable stepsizes, controlled for example by an LTE-based scheme, is often used in SPICE simulators to improve the runtime efficiency [74]. Using the Backward Euler method as an example, if the solution at t_n is computed, the next time step h_{n+1} is determined by

$$h_{n+1} = \sqrt{\frac{2\epsilon}{x''(\xi)}}, \tag{8.2}$$

where $h_{n+1} = t_{n+1} - t_n$, ϵ is the user-defined bound for LTE, and $x''(\xi)$ is estimated using the second-order divided difference $DD_2(t_n)$. After x_{n+1} is solved using h_{n+1}, the LTE formula in Equation (8.1) is checked to decide whether the computed x_{n+1} shall be accepted or not. If the LTE at t_{n+1} is within the given tolerance, x_{n+1} is accepted, otherwise, it is rejected and recomputed using a smaller stepsize.

Note that the use of TR tends to produce larger time steps than BE under the same LTE bound. However, TR may cause self-oscillation and for stiff circuits the stepsize may have to be reduced. In some cases, the numerical integration scheme may need to be switched from TR to the more robust BE.

8.1.2 Multistep Integration Methods

Compared to the two methods discussed earlier, different tradeoffs between speed and robustness are offered by Gear's methods [38]. The first and second-order Gear's methods are stiffly stable. Hence, they do not cause self-oscillation. Gear's methods fall into a family of multistep methods which rely on the circuit solutions at multiple preceding time points. The fixed stepsize second-order Gear's method (Gear2) is given by:

$$x_{n+1} = \frac{4}{3}x_n - \frac{1}{3}x_{n-1} + \frac{2}{3}h_{n+1}x'_{n+1}.$$

Variable stepsizes are preferred in practice, where the coefficients in the integration formula are determined dynamically. The variable stepsize Gear2 formula and its LTE are given in Equations (7.2) and (7.3), respectively.

8.1.3 Variable-Order Variable-Stepsize Integration Methods

In SPICE simulators, the first and second-order Gear's methods are widely adopted due to their stiff stability. Higher order Gear's methods are not commonly used.

However, there do exist other robust high order integration methods, some of which have been explored in MAPS. Despite their less familiarity to the CAD community, they have gained great success in numerical analysis and scientific computing. High order integration methods (order higher than two) have the potential advantages of producing large stepsizes. However, in practice it is well-known that high order methods may be unstable for certain ODEs. The potential instability of a high order method puts it at disadvantage. If a constant high order integration method, say fifth order, is used to integrate a stiff system, the stepsize may have to be reduced to be very small in order to maintain stability. Therefore, efficient high order integration methods usually have built-in mechanisms that dynamically vary both the order and the stepsize. DASSL is one of the most successful such variable-order variable-stepsize methods [9]. DASSL uses the fixed leading coefficient BDF formulas and is adopted in MAPS.

DASSL uses local truncation error to control the stepsize as well as the order. It estimates the local truncation error incurred at t_n if the step to x_n were taken at orders $k-2, k-1, k$, and $k+1$, respectively. Based on these error estimates, DASSL determines a chosen order k' for the next time step. If x_n is accepted, k' will be used to compute the solution at the future time point t_{n+1}; otherwise, k' will be used to re-compute x_n. The basic procedure of DASSL can be stated as:

(1) Compute the solution at t_n using predicted stepsize h_n, where $h_n = t_n - t_{n-1}$.
(2) Use the error estimates at t_n to decide order k' for the next step.
(3) Use the LTE at t_n to decide whether x_n should be accepted or re-computed.
(4) Predict the next stepsize: h_{n+1} if x_n is accepted; a new h_n if x_n is re-computed.

Readers may refer to [9] for a more complete discussion.

8.1.4 Exploiting Diversity in Numerical Integration

The foregoing discussion reveals that numerical integration methods vary in complexity, speed, and robustness. The one-step first-order BE method is robust, but has large LTEs. Variable-order variable-stepsize methods (e.g., DASSL) may produce much smaller LTEs and potentially lead to large stepsizes. However, they are significantly more complex and require numerous additional computations and checks to maintain accuracy and stability. For stiff circuits or stiff periods of the simulation, variable-order methods may decrease the integration order to one in order to ensure stability after attempting to stay at higher orders but failing in do so, which causes a large amount of rejected and wasted compute work.

In practice it is difficult to choose a single optimal numerical integration method for a simulation task *a priori*. The efficiency of a method is decided by the nature of the circuit and input stimulus, and it varies over the time as the circuit passes through various operation regimes. Integrating various numerical integration methods under MAPS allows the resulting algorithmic diversity be explored.

8.2 Algorithmic Diversity in Nonlinear Iterative Methods

Three nonlinear iterative methods have been attempted in the multi-algorithm simulation framework: the Newton's method, the successive chord (SC) method, and the secant method [81].

Despite its good robustness, the Newton's method is more expensive than the other two methods. At each Newton iteration, a new Jacobian matrix must be assembled, leading to expensive evaluation of device model derivatives. The new Jacobian matrix must be factorized, which is expensive for large circuits.

The SC method is a constant-Jacobian-type iterative method. The Jacobian matrix is either constructed once for the entire simulation or infrequently updated during the simulation. Therefore, the cost associated with assembling and factorizing the Jacobian matrix is significantly reduced. However, the convergence rate of the SC method is only linear and an infrequent update of the Jacobian matrix may lead to divergence during highly active phases of the circuit.

The secant method provides a different efficiency vs. complexity tradeoff compared with the above two methods. It forms a new Jacobian matrix at each iteration, but does so approximately by utilizing the information available from two consecutive time points. As such, the secant method reduces the cost of device model evaluation for stamping the Jacobian matrix. However, a new approximate Jacobian matrix must be factorized in each iteration. In addition, it has a superlinear convergence rate which is also inferior to the quadratic convergence rate of Newton's method. While the scant method seems to provide a comprise between Newton's method and SC method, practical experiences have shown that the inclusion of the secant method into multialgorithm simulation does not provide significant benefits. It was not selected in the current implementation of MAPS [120, 121].

It is worth mentioning that the nature of the multialgorithm simulation framework allows for unique exploration of nonrobust simulation algorithms. Nonrobust methods are seldom used as a standalone algorithm. In MAPS, since a robust Newton-based method is always chosen as a trustable backup, other nonrobust methods no longer have to converge during the entire simulation. MAPS extends the usability of risky algorithms for practical circuit simulation significantly by relaxing the need for being completely robust. In fact, it may be rather beneficial to purposely introduce *aggressive* or *risky* algorithms to opportunistically gain large runtime speedups during parts of the simulation, hence boosting the overall speedups of the parallel multialgorithm framework.

8.3 Interalgorithm Synchronization and Communication

To form a collaborative multialgorithm simulation framework, various simulation algorithms shall interact with each other via synchronization. Figure 8.2 depicts a conceptual synchronization scheme. The entire simulation period is divided into smaller time intervals. All the algorithms synchronize with each other at the end of each interval: (1) the first algorithm that reaches the end of an interval is deemed the *winner* for the interval; (2) the winner informs other algorithms that they have fallen behind; (3) the algorithms that fell behind quit from their current simulation loop and jumpstart from the end of the

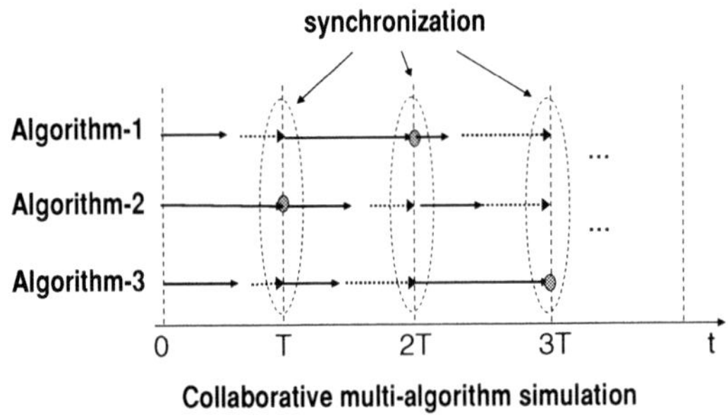

Fig. 8.2 The conceptual multialgorithm synchronization scheme.

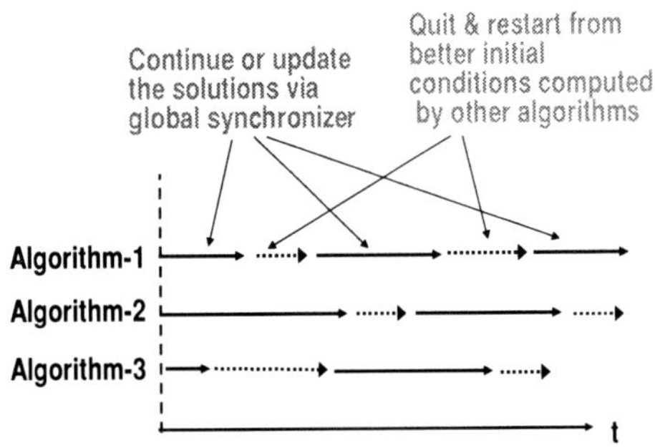

Fig. 8.3 A more flexible multialgorithm synchronization scheme (from [121] ©[2011] IEEE).

interval (the start of the next interval) using the work computed by the winner algorithm as the initial condition.

A more flexible and cleaner synchronization model is shown in Figure 8.3. In this case, the algorithms do not directly interact with each other, rather, they *asynchronously*, or *independently*, communicate with a *global synchronizer*. The global synchronizer stores the circuit solutions at the k most recent time points, where k is determined by the highest order of numerical integration used. This scheme allows the communication granularity to be controlled on an individual algorithm basis and provides a more transparent interface between multiple algorithms.

Essentially, what the global synchronizer provides is a coarse-grained communication scheme which enables solution sharing between algorithms. Algorithmic diversity is naturally exploited when a well performing algorithm writes its latest solution into the global synchronizer.

Every algorithm reads the global synchronizer to get the latest solutions as its initial condition for future time points. It is also possible to share the factorized Jacobian matrices computed by the Newton's method with the SC method. This "free" update of the Jacobian can lead to improved convergence for the SC method [121].

In the current implementation, the three numerical integration methods with independent dynamic stepsize control were all paired with Newton's method to form three complete simulation algorithms. Among these, the SPICE-like combination's of the BE and Newton's methods was always selected, which provided a basic guarantee for the success of the simulation. BE was also paired with the successive chord to create the fourth algorithm.

As a coarse-grained parallel approach, the multialgorithm framework is completely orthogonal to the actual implementations of individual algorithms. When sufficient hardware sources are available, each algorithm can be further parallelized by using any form of intraalgorithm parallelism. Parallel device model evaluation and matrix solution were adopted to further parallelize each algorithm to form a two-level hierarchical MAPS framework on shared-memory machines in [121], as shown in Figure 8.4.

The four-algorithm MAPS was implemented using four threads on a shared-memory Linux server with two Intel Xeon quad-core 2.33 GHZ processors. Compared with the serial BE + Newton-based simulation, for a set of test circuits, the parallel speedups achieved by MAPS are shown in Figure 8.5. Note that for some test circuits, the speedup was much greater than four times for this four-thread MAPS implementation. These large speedups were resulted from the multialgorithm collaborations between the various algorithms, unique for MAPS. An eight-thread HMAPS implementation, where each of the four algorithms was further parallelized using two-thread-based parallel device model evaluation and matrix solve, was demonstrated in [121]. For larger circuits, this HMAPS implementation introduced an additional speedup of a factor of two.

The total runtime of the four-thread MAPS is broken down to synchronization and computational costs for several circuits in Figure 8.6. It is clearly seen that the synchronization of multialgorithm simulation

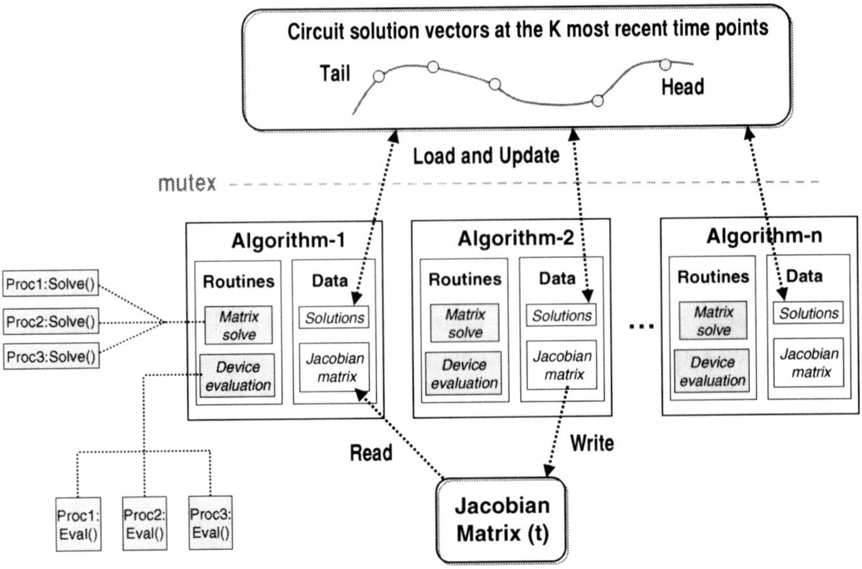

Fig. 8.4 Two-level hierarchical multialgorithm simulation (HMAPS) (from [121] ©[2011] IEEE).

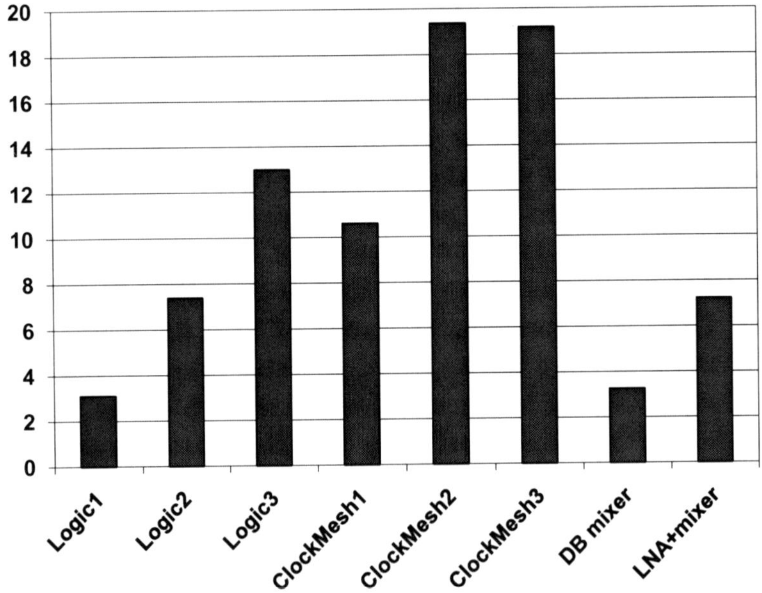

Fig. 8.5 Speedups of a four-algorithm four-thread MAPS implementation on a shared-memory machine.

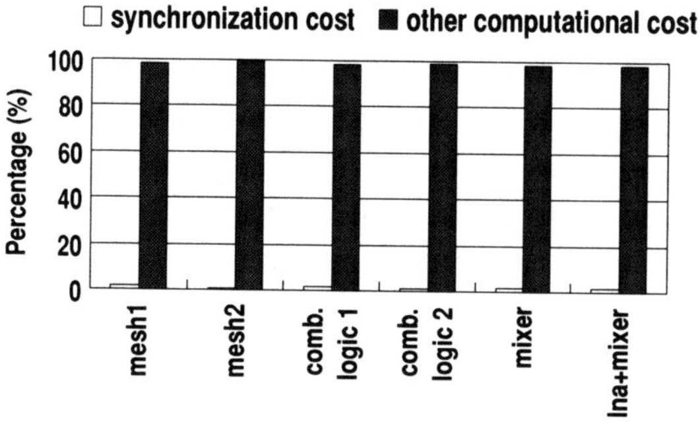

Fig. 8.6 Synchronization cost vs. computational cost for the four-thread MAPS implementation (from [121] ©[2011] IEEE).

is very low due to the coarse-grained nature of the interalgorithm parallelism that was explored.

8.4 Parallel Performance Modeling and Runtime Optimization

While exploiting a combination of inter- and intra-algorithm parallelism leads to good performance scalability in HMAPS, flexibilities exist when choosing a subset of active algorithms from an algorithm tool and allocating parallel resources (e.g., number of threads) for each active algorithm. Such flexibilities lead to a large configuration space for HMAPS. Experimentally, large runtime variations (about one order of magnitude) have been observed across different HMAPS configurations [122, 123]. Selecting an optimal configuration is completely nontrivial as the parallel simulation efficiency critically depends on the size and nature of the circuit being simulated, parallel scalability of various parts of the code, complex hardware characteristics, multialgorithm interactions, and so on.

To optimize the efficiency of HMAPS statically, or preruntime, a static parallel performance modeling approach was developed in [123]. A decomposable approach was taken to characterize the runtime performances of key components of a simulation algorithm and interalgorithm collaboration as functions of design characteristics. The resulting set of models were used to efficiently predict the runtimes of all possible HMAPS configurations such that the best configuration was chosen prior to the start of the simulation.

The above approach was further extended to allow for inclusion of on-the-fly runtime data to the static performance models. It makes it possible to dynamically reconfigure HMAPS for a given circuit being simulated based upon valuable runtime information gathered as the simulation proceeds, contributing to additional performance boost of HMAPS [122].

9

Elements of Fast-SPICE Simulation Techniques

While the main focus of this monograph is on parallel transistor-level simulation techniques that deliver full SPICE-level accuracy, it is of an interest to briefly present key elements of the recent fast-SPICE development. For this monograph, doing so may be highly relevant as ideas that are behind some of the techniques reviewed in previous sections (e.g., nonlinear relaxation and waveform relaxation methods) have successfully evolved into fast-SPICE simulators. Nevertheless, it shall be noted that fast-SPICE exploits a large variety of techniques, only a subset of which are approached from a parallel computing perspective.

The unsaturated need for simulating increasingly complex integrated circuits has stimulated the development of fast-SPICE where it may become meaningful to trade off simulation accuracy for large performance and capacity gains. Large-scale digital, mixed-signal, memory circuit blocks, or full-chip designs may have more than 100 million MOSFET and parasitic devices. The simulation of such large designs with full SPICE-level accuracy can become computationally intractable. There are cases where designers are willing to pay the price of 5–15% accuracy loss for 1000-fold or more runtime speedups. This opens up opportunities for the designers to trade off some accuracy for speed gain in order to make some of the extremely challenging large-scale simulations feasible. Fast-SPICE simulators achieve this goal by employing a myriad of techniques, heuristics, and methodologies. In the following, several key fast-SPICE simulation techniques are briefly discussed. More detailed discussions of the subject can be found from several sources [14, 26, 32, 108] and a recent survey [91].

9.1 Techniques Stimulated by the Development of Fast-SPICE

There are several techniques that are not unique to fast-SPICE but have been directly or indirectly stimulated by the development of fast-SPICE. Among these, efficient preconditioned and iterative matrix solution methods have found their applications in fast-SPICE simulators, in which distributed and multithreaded matrix solvers may also be adopted. As seen from the previous sections, these techniques are not fast-SPICE-specific and have been exploited in accurate SPICE-level simulation applications as well.

In addition, on-the-fly code generation and compilation is commonly employed in fast-SICE for efficient evaluation of key circuit blocks. Finally, efficient implementation of the *front end* of the simulator is critical for highly hierarchical designs such as memories. Proper representation and storage of the input data and hierarchical netlists are shown to be important for setting up the simulation and processing the simulation data.

9.2 Approximate Device and Parasitics Models

As stated in Section 3, device model (e.g., MOSFET models) evaluation may contribute significantly to the cost of circuit simulation. In fast-SPICE simulators, accurate equation-based device models may be replaced by simplified equation models, or, much more often, table lookup models. For the latter, the current, capacitance and charge characteristics of MOSFETs are precharacterized and tabulated for fast evaluation. In the simulation of mixed-signal designs, approximated device models are typically only applied to parts of the circuit that operate in the digital regime.

Along a similar line, approximate models are also employed for parasitics for the simulation of large fully extracted designs, where linear RLC nets can become a significant part of the netlist. A proper reduction of parasitics can speed up the simulation rather dramatically. To this end, projection-based Krylov-subspace model reduction algorithms [79], a popular choice for parasitics reduction, compute a smaller reduced model by matching a certain number of moments of the original transfer functions. The error of the reduced-order model can be practically reduced by increasing the number of matched moments. To alleviate some of the challenges in reducing linear parasitic network with many input/output ports, heuristics and graph-based reduction algorithms have also emerged [99].

9.3 Event-Driven and Multirate Simulation

The standard (accurate) transient circuit simulation techniques are developed by forcing global convergence and globally uniform time-stepping. In contrast, fast-SPICE relaxes the two requirements so as to allow for efficient event-driven multirate simulation.

To this end, a circuit is partitioned into several loosely coupled subnetworks [97]. The numerical simulation of a subnetwork can be conducted based upon a different sequence of time points. This allows each subnetwork to be simulated with stepsizes that are tailored for the rate of change of the subnetwork, thereby facilitating multirate simulation of the entire network. Equally importantly, at any time, a subnetwork is simulated if and only if its state is expected to change due to an appreciative input or load change. And if this is the case, the subnetwork is considered to be *alert*. Otherwise, the subnetwork is deemed dormant and full simulation is skipped.

In event-driven multirate simulation, the entire network is not simulated as a fully coupled system. In other words, no global convergence is targeted and only local convergence is achieved while numerically solving individual subnetworks, which is done without a globally uniform time stepping for the complete circuit. It is possible to simultaneously process multiple events, leading to parallelization of event-driven simulation.

It is worth mentioning that some of the philosophies behind the event-driven multirate simulation discussed above have also been embraced in the nonlinear relaxation [18, 76] and waveform relaxation methods [65, 118] introduced in Section 4, but in different ways. The nonlinear relaxation and waveform relaxation methods explore Gauss–Jacobi or Gauss–Seidel-type relaxations at the nonlinear equation and differential equation levels, respectively.

The event-driven multirate technique discussed here, on the other hand, centers more on the detection of circuit events or activities, and efficient processing of these events once detected. Moreover, over the years, as part of the development of fast-SPICE technologies, more sophisticated circuit partitioning schemes have emerged [60]. These techniques go beyond partitioning the circuit by grouping channel connected MOSFETs and are also capable of identifying different types of circuit blocks (e.g., digital/analog circuits and power supply networks), hence producing more intelligent partitions. Such partitioning techniques become effective enablers for event-driven multirate fast-SPICE simulations.

9.4 Hierarchical Isomorphic Simulation

Hierarchical isomorphic simulation, which is exploited in typical fast-SPICE class simulators (cf. e.g., [61, 92, 108]), comes with a rather different flavor compared with the parallel simulation techniques that are introduced in previous sections. The key motivating observation here is that many circuits employ repetitive structures. For example, the same standard cells may be populated many times in digital designs and the same memory cell is reused millions of time to create large memory arrays. In mixed-signal designs, same basic circuit topologies may be reused many times. To allow for more efficient simulation of such designs in terms of both simulation time and memory usage, only one or several representative circuit instances of the same topology are simulated under identical or similar input stimuli.

In particular, in the case of accelerated memory simulation, techniques have been developed to identify and remove latent parts of the memory array from the simulation based on a hierarchical representation of the memory circuit. More recent approaches are able to automatically detect several cell and memory topologies and exploit typical operating patterns. By doing so, a memory array of N cells is modeled efficiently by $\sqrt{N}$ row or column models, yielding significant reduction [61, 91]. Techniques targeting more accurate consideration of layout parasitics in memory simulation have also been developed [92].

10

Parallel Circuit Simulation on Heterogeneous Platforms

Accelerating circuit simulation using special hardware acceleration is not a completely new concept. However, the recent emergence of high-performance heterogeneous platforms such as the Cell processor [86] and computer systems with attached GPU [4, 78] and FPGA accelerators has attracted new interest in developing hardware-accelerated circuit simulation techniques. While these hardware platforms may offer impressive computing power, some of their architectural limitations must be dealt with when developing general-purpose applications such as circuit simulation.

Hardware acceleration for circuit simulation has primarily focused on accelerating two most time-consuming components in the simulation flow: transistor model evaluation and sparse matrix solution. Some of the recent activities will be briefly reviewed.

In a broader context, we will also comment on the ongoing and future trends toward leveraging heterogeneous platforms that integrate general-purpose, SIMD processors and dedicated accelerators, operating with both the shared and distributed memory models. We will outline several key new research needs in this direction.

10.1 Acceleration of Device Model Evaluation

The fact that transistor model evaluations are completely independent of each other has made massively parallel SIMD GPUs an appropriate choice for acceleration. The acceleration of BSIM3 device model on GPU using the CUDA programming interface was demonstrated in [43]. To maximize the runtime efficiency, inlining of if-then-else statements and optimal partitioning of the BSIM3 code into CUDA kernels of appropriate sizes were performed. Optimized usage of memory resources including global memory, texture memory, and constant memory was focused to facilitate fast memory access and latency hiding. The GPU-based model evaluation was interfaced with a commercial simulator, leading to an average speedup of 2.36 times for accelerated circuit simulation. GPU-based acceleration of BSIM4 device model evaluation was demonstrated in [88], resulting in a speedup factor of 3 to 6 for circuit simulation under a different simulation environment.

An automated code generation and performance-tuning approach for evaluation of Verilog-AMS-based transistor models on FPGA, GPU, cell, and multicore processors was presented in [55]. Auto tuning was done by exploring several implementation parameters for different hardware architectures and picking the parameter setting that delivers the best performance. Model evaluation speedup factors of 3–182 for a Xilinx Virtex5 LX330T, 1.3–33 for a Cell processor, and 3–130 for an NVIDIA 9600 GT GPU over the execution of device model evaluation on a 3 GHz Intel Xeon 5160 processor were demonstrated.

Compared to other processor platforms, reconfigurable FPGAs have a unique advantage. An FPGA can be reconfigured specifically for a parallel application. In [53], for SPICE model evaluation, a VLIW architecture is used. Different numbers of add, multiply, divide, square-root, and exponential operators are mixed in a processing element in proportion to their frequency of usage. Loop unrolling and software pipelining are used to optimize the performance.

10.2 Acceleration of Matrix or System Solution

An approach to parallelization of sparse matrix solve for circuit simulation on FPGA was presented in [54]. The KLU direct solver

[75], which maintains the same matrix structure from one Newton iteration to another, was parallelized on a 250 MHz Xilinx Virtex-5 FPGA, resulting speedups of 1.2–64 times over the solver code running on an Intel Core i7 965 processor. The development of dense linear algebra solvers on hybrid architectures, namely multicore processors with GPU accelerators, was presented in [107].

Accelerating large sparse matrix problems on GPU presents new challenges as standard algorithms significantly rely on random data structures and flow control, whose implementation on SIMD GPU platforms would lead to severe thread divergence and hence practically a slowdown of the execution. Recent efforts along this line have been on developing custom solvers [28, 29, 30, 69, 100, 129] or parallelizing sparse-matrix multiplications (an ingredient of an iterative solver) [17, 37]. Among these, GPU-accelerated multigrid solvers were developed for power delivery network simulation in [28, 30, 129], and IC thermal analysis in [29]. A two-dimensional Poisson equation-based formulation for analyzing a restricted class of regular on-chip power grids was developed in [100].

In [28], the challenges of accelerating large-scale irregular on-chip power grids on GPU were addressed by a combination of algorithmic and implementation-level considerations. This leads to a hybrid solver that involves iterations between the GPU accelerator and the CPU host in a manner that the "hidden" data parallelism in the application is properly exposed and exploited on GPU.

Typical industrial power grid designs exhibit a high degree of wiring regularity while local irregularity does exist, which is difficult to be handled on an SIMD GPU. In the approach of [28], a mostly regular multilayer power grid was first approximated by a fully regular grid. This regular grid approximate was solved efficiently by a custom multigrid solver on GPU. To leverage the SIMD compute power of GPU in a highly efficient way, the multigrid solver was designed to operate on regular control flows using custom regular data structures. To ensure the overall analysis accuracy, the solution returned from the GPU was checked on the CPU host based on the original power grid; this checking processing was computationally straightforward and could be done easily on the CPU by leveraging its flexibility in processing irregular data structure and flows as required for processing the original irregular grid. The computed residue was returned to the GPU solver to compute the correction to the solution. A few such GPU–CPU approximate-and-correct iterative steps were performed to arrive at a fully converged solution.

As such, the strengths of both general-purpose CPU and SIMD GPU hardware were exploited simultaneously, leading to the overall high-efficiency of the solver. Up to 50X speedups over the pure CPU-based implementation of the same algorithm have been demonstrated by the hybrid solver for a set of industrial power grid benchmarks [49]. The approach was extended to a multicore-multi-GPU platform where additional parallel gains were demonstrated [30]. It is also possible to further enhance the simulation robustness by adopting this GPU–CPU hybrid solver as a preconditioner for a top-level iterative method.

Arithmetic intensity, a critical factor that measures the amount of computation per global memory access, shall be maximized in order to achieve efficient GPU progressing. The work in [28] increases arithmetic intensity by performing multiple point-wise relaxation iterations on each loaded circuit block. This efficiently turns the numerical relaxation scheme used in the multigrid algorithm into a global block Gauss–Siedel method.

The GPU-accelerated on-chip power grid solver was extended to analyzing large power delivery networks consisting of off-chip (package and PCB) circuitry, on-chip power grids with multiple voltage domains and integrated nonlinear voltage regulators in [129]. A partitioning-based strategy was adopted where large on-chip power grids were efficiently processed on the GPU while the off-chip circuitry and active regulator circuits were processed by a SPICE-like simulation engine running on the CPU host. The efficiency of this simulation infrastructure was leveraged to optimize large power delivery networks with a large number of on-chip voltage regulators [129].

10.3 Ongoing and Future Directions

Many parallel simulation techniques that have been discussed so far were developed on distributed, multicore, or shared-memory machines, and/or hardware accelerators. The current technology trend is moving toward providing affordable parallel computing platforms with even a greater performance to the designers. The emergence of high-performance computer clusters with networked heterogeneous multi-/many-core processors and accelerators is putting more parallel compute power at the disposal of the designers for tackling circuit simulation challenges. This provides new challenges and opportunities.

A holistic framework for developing parallel circuit simulation techniques is very much needed in order to fully explore the power of such massively parallel systems. It is expected that a multiplicity of forms of parallelism, possibly orthogonal to each other, needs to be leveraged on general-purpose, SIMD processors and dedicated accelerators, on packaged processor dies, in shared-memory nodes, or across the network. A plethora of parallel simulation techniques across all algorithmic levels (Figure 2.2) and in several domains of parallel processing (Figure 2.3) may be targeted.

Key architecture properties including heterogeneity of processors, core-to-core and core-to-memory bandwidths, communication/synchronization overheads within the die/node and across the cluster must be considered throughout the algorithm design phase. Specific algorithmic and implementation decisions must be evaluated with a good understanding of their interactions with the underlying architecture. Circuit-simulation-specific parallel performance models and code tuning capabilities may be developed for guiding the exploration in this complex design and optimization space.

Finally, this holistic development framework may also stimulate the development of "new" parallel algorithms as opposed to the parallelization of existing serial algorithms. While the boundary between the two is not always clearly defined, there do exist new opportunities for creating novel paradigms as the landscape of parallel computing continues to evolve.

11

Conclusions

Accurate, robust, and fast transistor-level circuit simulation has important implications on the design and verification of integrated circuits and systems. Parallel circuit simulation has drawn significant research interest both historically and at the present time. Parallel circuit simulation will remain as an active area of research as the proliferation of parallel computing architectures continues. To move forward, we expect both algorithmic innovation and efficient implementation are of critical importance. In particular, we envision that a holistic approach that captures interactions between algorithmic design, implementation strategies, and hardware architectures will result in fruitful outcomes and enable a full leverage of growing compute power that is predominantly available in the form of parallelism. The ability in utilizing emerging homogenous and heterogeneous parallel architectures will open up new opportunities in advancing circuit simulation.

Acknowledgments

The author would like to thank many colleagues and collaborations who have contributed either directly or indirectly to the materials presented in this monograph. In particular, the author wants to thank his current and former students, Wei Dong, Xiaoji Ye, Albert (Zhiyu) Zeng, and Zhuo Feng for their contributions to several parallel circuit techniques that are discussed in this paper. The author thanks Sani Nassif for collaborating on multialgorithm parallel circuit simulation. The author is grateful to Lawrence Pileggi, Lawrence Arledge, Peter Fang, Kiran Gullapalli, Frank Liu, Ben Gu, and Emrah Acar for their support and interactions in the past many years, which helped the author gain a deep understanding of circuit simulation. The assistance and patience received from Sharad Malik and James Finlay during the preparation of this manuscript is gratefully acknowledged.

This material is based upon work supported by the National Science Foundation under grants numbers 0747423 and 0903485, the Semiconductor Research Corporation and the Texas Analog Center for Excellence under Contract 2008-HC-1836, and the Semiconductor Research Corporation under Contract 2009-TJ-1987.

References

[1] *Accelerating Analog Simulation with HSPICE Precision Parallel Technology,* White Paper, Synopsys, Inc., http://www.synopsys.com/Tools/Verification/AMSVerification/CircuitSimulation/HSPICE/CapsuleModule/HSPICE-HPP-wp.pdf, 2011.

[2] AMD Fusion APUs, http://sites.amd.com/us/Documents/48423B_fusion_whitepaper_WEB.pdf.

[3] AMD Opteron™ dual-core processors, http://www.amd.com/us-en/Processors/ProductInformation/0,,30_118_8796_14309,00.html.

[4] AMD Radeon, http://www.amd.com/us/products/desktop/graphics/amd-radeon-hd-6000/hd-6990/Pages/amd-radeon-hd-6990-overview.aspx.

[5] G. Antonini and A. E. Ruehli, "Waveform relaxation time domain solver for subsystem arrays," *IEEE Transactions on Advanced Packaging*, vol. 33, no. 4, pp. 760–768, November 2010.

[6] K. Asanovic, R. Bodik, B. Catanzaro, J. Gebis, P. Husbands, K. Keutzer, D. Patterson, W. Plishker, J. Shalf, S. Williams, and K. Yelick, "The landscape of parallel computing research: A view from Berkeley," Technical Report UCB/EECS-2006-183, EECS Department, University of California, Berkeley, http://www.eecs.berkeley.edu/Pubs/TechRpts/2006/EECS-2006-183.html, December 2006.

[7] A. Bataineh, M. Aamodt, and K. Thomas, "A parallel and vector implementation of circuit simulation on cray supercomputers," *Parallel Algorithms Applications*, vol. 14, no. 2, pp. 109–118, 1999.

[8] S. Borkar, "Thousand core chips — a technology perspective," in *IEEE/ACM Design Automation Conference*, pp. 746–749, June 2007.

[9] K. E. Brenan, S. L. Campbell, and L. R. Petzold, *Numerical Solutions of Initial-Value Problems in Differential-Algebraic Equations.* New York, USA: Elsevier Science Publishing Co., Inc., 1989.

[10] Cadence Spectre, `http://www.cadence.com/products/cic/spectre_circuit`.

[11] B. Catanzaro, K. Keutzer, and B. Su, "Parallelizing cad: A timely research agenda for eda," in *ACM/IEEE Design Automation Conference (DAC 2008)*, pp. 12–17, June 2008.

[12] M.-C. Chang and I. N. Hajj, "iPRIDE: A parallel integrated circuit simulator using direct method," in *IEEE International Conference on Computer-Aided Design (ICCAD-88), Digest of Technical Papers*, pp. 304–307, November 1988.

[13] P. F. Cox, R. G. Burch, D. E. Hocevar, P. Yang, and B. D. Epler, "Direct circuit simulation algorithms for parallel processing [VLSI]," *IEEE Transactions on Computer-Aided Design of Integrated Circuits and Systems*, vol. 10, no. 6, pp. 714–725, June 1991.

[14] *CustomSim Datasheet,* Synposys, Inc., `http://www.synopsys.com/Tools/Verification/AMSVerification/CircuitSimulation/Documents/CustomSim-ds.pdf`.

[15] F. Dartu and L. T. Pileggi, "Teta: Transistor-level engine for timing analysis," in *Proceedings of the Design Automation Conference*, pp. 595–598, June 1998.

[16] J. W. Demmel, J. R. Gilbert, and X. S. Li, "An asynchronous parallel supernodal algorithm for sparse Gaussian elimination," *SIAM Journal on Matrix Analysis and Applications*, vol. 20, no. 4, pp. 915–952, 1999.

[17] Y. Deng, B. D. Wang, and S. Mu, "Taming irregular EDA applications on GPUs," in *IEEE/ACM International Conference on Computer-Aided Design–Digest of Technical Papers (ICCAD 2009)*, pp. 539–546, November 2009.

[18] J. T. Deutschand and A. R. Newton, "A multiprocessor implementation of relaxation-based electrical circuit simulation," in *21st Conference on Design Automation*, pp. 350–357, June 1984.

[19] A. Devgan and R. A. Rohrer, "Adaptively controlled explicit simulation," *IEEE Transactions on CAD*, vol. 13, no. 6, pp. 746–762, June 1994.

[20] W. Dong and P. Li, "Accelerating harmonic balance simulation using efficient parallelizable hierarchical preconditioning," in *ACM/IEEE Design Automation Conference (DAC '07)*, pp. 436–439, June 2007.

[21] W. Dong and P. Li, "Hierarchical harmonic-balance methods for frequency-domain analog-circuit analysis," *IEEE Transactions on Computer-Aided Design of Integrated Circuits and Systems*, vol. 26, no. 12, pp. 2089–2101, December 2007.

[22] W. Dong and P. Li, "Parallelizable stable explicit numerical integration for efficient circuit simulation," in *ACM/IEEE Design Automation Conference (DAC '09)*, pp. 382–385, July 2009.

[23] W. Dong and P. Li, "A parallel harmonic-balance approach to steady-state and envelope-following simulation of driven and autonomous circuits," *IEEE Transactions on Computer-Aided Design of Integrated Circuits and Systems*, vol. 28, no. 4, pp. 490–501, April 2009.

[24] W. Dong, P. Li, and X. Ye, "Wavepipe: Parallel transient simulation of analog and digital circuits on multi-core shared-memory machines," in *ACM/IEEE Design Automation Conference (DAC 2008)*, pp. 238–243, June 2008.

[25] J. Dorsey, S. Searles, M. Ciraula, S. Johnson, N. Bujanos, D. Wu, M. Braganza, S. Meyers, E. Fang, and R. Kumar, "An integrated quad-core Opteron^{TM} processor," in *IEEE International Solid-State Circuits Conference (ISSCC)*, pp. 102–103, February 2007.

[26] *FastSPICE simulation in the analog design environment,* White Paper, Cadence Design Systems, Inc., `http://www.cadence.com/rl/Resources/white_papers/5933_FastSPICEsim_WPfnl.pdf`, 2004.

[27] P. Feldmann, B. Melville, and D. Long, "Efficient frequency domain analysis of large nonlinear analog circuits," in *Proceedings of the IEEE 1996 Custom Integrated Circuits Conference*, pp. 461–464, May 1996.

[28] Z. Feng and P. Li, "Multigrid on GPU: Tackling power grid analysis on parallel simt platforms," in *IEEE/ACM International Conference on Computer-Aided Design (ICCAD 2008)*, pp. 647–654, November 2008.

[29] Z. Feng and P. Li, "Fast thermal analysis on GPU for 3D-ICS with integrated microchannel cooling," in *IEEE/ACM International Conference on Computer-Aided Design (ICCAD 2010)*, pp. 551–555, November 2010.

[30] Z. Feng, Z. Zeng, and P. Li, "Parallel on-chip power distribution network analysis on multi-core-multi-GPU platforms," *IEEE Transactions on Very Large Scale Integration (VLSI) Systems*, vol. 19, no. 10, pp. 1823–1836, October 2011.

[31] C. M. Fiduccia and R. M. Mattheyses, "A linear-time heuristic for improving network partitions," in *Conference on Design Automation*, pp. 175–181, June 1982.

[32] *FineSimPro Datasheet,* MAGMA Design Automation, Inc., `http://www.magma-da.com/uploadedFiles/products-solutions/analysis/FineSim-Pro%20Datasheet.pdf`, 2008.

[33] M. J. Flynn, "Very high-speed computing systems," *Proceedings of the IEEE*, vol. 54, no. 12, pp. 1901–1909, December 1966.

[34] J. Friedrich, B. McCredie, N. James, B. Huott, B. Curran, E. Fluhr, G. Mittal, E. Chan, Y. Chan, D. Plass, H. L. S. Chu, L. Clark, J. Ripley, S. Taylor, J. Dilullo, and M. Lanzerotti, "Design of the Power6^{TM} microprocessor," in *IEEE International Solid-State Circuits Conference (ISSCC)*, pp. 96–97, February 2007.

[35] N. Frohlich, B. M. Riess, U. A. Wever, and Q. Zheng, "A new approach for parallel simulation of VLSI circuits on a transistor level," *IEEE Transactions on Circuits and Systems I: Fundamental Theory and Applications*, vol. 45, no. 6, pp. 601–613, June 1998.

[36] U. Gajanan, M. Hassan, L. Warriner, K. Yen, B. Upputuri, D. Greenhill, A. Kumar, and H. Park, "An 8-core 64-thread 64b power-efficient SPARC, SoC," in *IEEE International Solid-State Circuits Conference (ISSCC)*, pp. 108–109, February 2007.

[37] M. Garland, "Sparse matrix computations on manycore GPUs," in *ACM/IEEE Design Automation Conference (DAC 2008)*, pp. 2–6, June 2008.

[38] C. W. Gear, *Numerical Initial Value Problems in Ordinary Differential Equations.* New Jersey, USA: Prentice-Hall, Englewood Cliffs, 1971.

[39] C. W. Gear and I. G. Kevrekidis, "Projective methods for stiff differential equations: Problems with gaps in their eigenvalue spectrum," *SIAM Journal on Scientific Computing*, vol. 24, no. 4, pp. 1091–1106, 2002.

[40] C. W. Gear and I. G. Kevrekidis, "Telescopic projective methods for parabolic differential equations," *Journal of Commotional Physics*, vol. 187, no. 1, pp. 95–109, 2003.

[41] C. W. Gear and I. G. Kevrekidis, "Computing in the past with forward integration," *Physics Letters A*, vol. 321, pp. 335–343, 2004.

[42] R. Griffith and M. Nakhla, "A new high-order absolutely-stable explicit numerical integration algorithm for the time-domain simulation of nonlinear circuits," in *IEEE/ACM International Conference on Computer-Aided Design, Digest of Technical Papers*, pp. 276–280, November 1997.

[43] K. Gulati, J. F. Croix, S. P. Khatri, and R. Shastry, "Fast circuit simulation on graphics processing units," in *Asia and South Pacific Design Automation Conference (ASP-DAC 2009)*, pp. 403–408, January 2009.

[44] A. Gupta, "Parallel sparse direct methods: A short tutorial," *IBM Research Report*, RC, (25076), 2010.

[45] J. Held, J. Bautisa, and S. Koehl, "From a few cores to many: A tera-scale computing research overview," Intel Research White Paper, `http://download.intel.com/research/platform/terascale/terascale_overview_paper.pdf`, 2006.

[46] M. Honkala, J. Roos, and M. Valtonen, "New multilevel Newton-Raphson method for parallel circuit simulation," in *Proceedings of European Conference on Circuit Theory and Design*, pp. II-113–II-116, August 2001.

[47] G. G. Hung, K. Gallivan, and R. Saleh, "Parallel circuit simulation based on nonlinear relaxation methods," in *IEEE International Sympoisum on Circuits and Systems*, vol. 4, pp. 2284–2287, June 1991.

[48] G. G. Hung, Y. C. Wen, K. Gallivan, and R. Saleh, "Parallel circuit simulation using hierarchical relaxation," in *Proceedings of the ACM/IEEE Design Automation Conference*, pp. 394–399, June 1990.

[49] IBM power grid benchmark suite, `http://dropzone.tamu.edu/~pli/PGBench/`.

[50] Intel Core^{TM} 2 Qual processors, `http://www.intel.com/products/processor/core2quad/index.htm`.

[51] G. K. Jacob, A. R. Newton, and D. O. Pederson, "An empirical analysis of the performance of a multiprocessor-based circuit simulator," in *23rd Conference on Design Automation*, pp. 588–593, June 1986.

[52] T. A. Johnson and A. E. Ruehli, "Parallel waveform relaxation of circuits with global feedback loops," in *Proceedings of the ACM/IEEE Design Automation Conference*, pp. 12–15, June 1992.

[53] N. Kapre and A. DeHon, "Accelerating spice model-evaluation using FPGAS," in *IEEE Symposium on Field Programmable Custom Computing Machines (FCCM '09)*, pp. 37–44, April 2009.

[54] N. Kapre and A. DeHon, "Parallelizing sparse matrix solve for SPICE circuit simulation using FPGAs," in *International Conference on Field-Programmable Technology (FPT 2009)*, pp. 190–198, December 2009.

[55] N. Kapre and A. DeHon, "Performance comparison of single-precision spice model-evaluation on FPGA, GPU, Cell, and multi-core processors," in *International Conference on Field Programmable Logic and Applications (FPL 2009)*, pp. 65–72, August 2009.

[56] V. Karanko and M. Honkala, "A parallel harmonic balance simulator for shared memory multicomputers," in *European Microwave Conference*, pp. 849–851, October 2004.

[57] G. Karypis and V. Kumar, "hMETIS: A hypergraph partitioning package (version 1.5.3)," `http://glaros.dtc.umn.edu/gkhome/fetch/sw/hmetis/manual.pdf`.

[58] E. Keiter, R. Schiek, and H. Thornquist, "Parallel algorithm strategies for circuit simulation," *Research Report, Sandia National Laboratories*, SAND2010-0248:1–21, January 2010.

[59] B. W. Kernighan and S. Lin, "An efficient heuristic procedure of partitioning graphs," *Bell System Technical Journal*, vol. 49, pp. 291–307, February 1970.

[60] J. K. Kerns, M. Bhattacharya, S. Rudnaya, and K. Gullapalli, "Automatic hierarchy-independent partitioning method for transistor-level circuit simulation," United States Patent Application Publication, January 2009.

[61] J. K. Kerns and Z. Peng, "Spice optimized for arrays," United States Patent Application Publication, January 2008.

[62] E. S. Kuh and R. A. Rohrer, "The state-variable approach to network analysis," *Proceedings of the IEEE*, vol. 53, no. 7, pp. 672–686, July 1965.

[63] K. Kundert, J. White, and A. Sangiovanni-Vincentelli, *Steady-State Methods for Simulating Analog and Microwave Circuits.* Boston: Kluwer Academic Publisher, 1990.

[64] N. A. Kurd, S. Bhamidipati, C. Mozak, J. L. Miller, T. M. Wilson, M. Nemani, and M. Chowdhury, "Westmere: A family of 32 nm IA processors," in *IEEE International Solid-State Circuits Conference Digest of Technical Papers (ISSCC)*, pp. 96–97, February 2010.

[65] E. Lelarasmee, A. E. Ruehli, and A. L. Sangiovanni-Vincentelli, "The waveform relaxation method for time-domain analysis of large scale integrated circuits," *IEEE Transactions on Computer-Aided Design of Integrated Circuits and Systems*, vol. 1, no. 3, pp. 131–145, July 1982.

[66] P. Li and L. T. Pileggi, "Efficient harmonic balance simulation using multi-level frequency decomposition," in *IEEE/ACM International Conference on Computer Aided Design (ICCAD-2004)*, pp. 677–682, November 2004.

[67] P. Li, L. M. Silveira, and P. Feldmann, eds., *Simulation and Verification of Electronic and Biological Systems.* Springer, 2011.

[68] C. Lin and L. Synder, *Principles of Parallel Programming.* Addison Wesley, 2008.

[69] X. Liu, H. Yu, J. Relles, and S. X.-D. Tan, "A structured parallel periodic arnoldi shooting algorithm for RF-PSS analysis based on gpu platforms," in *Asia and South Pacific Design Automation Conference (ASP-DAC)*, pp. 13–18, January 2011.

[70] H. W. Mathers, S. R. Sedore, and J. R. Seuts, "Automated digital computer program for determining response of electronic circuits to transient nuclear

radiation (SCEPTRE)," Technical Report, IBM Federal Systems Division, Electronics Systems Center, Owego, NY, February, 1967.

[71] W. McCalla and D. Pederson, "Elements of computer-aided circuit analysis," *IEEE Transactions on Circuit Theory*, vol. 18, no. 1, pp. 14–26, January 1971.

[72] C. McNairy and R. Bhatia, "Montecito: A dua-core, dual-thread Itanium processor," *IEEE Micro*, vol. 25, no. 2, pp. 10–20, March 2005.

[73] L. Nagel and R. Rohrer, "Computer analysis of nonlinear circuits, excluding radiation (CANCER)," *IEEE Journal of Solid-State Circuits*, vol. 6, no. 4, pp. 166–182, August 1971.

[74] L. W. Nagel, *SPICE2: A Computer Program to Simulate Semiconductor Circuits*. ERL-M520, University of California, Berkeley, CA, USA, 1975.

[75] E. Natarajan, "KLU-a high performance sparse linear solver for circuit simulation problems," Master's Thesis, University of Florida, 2005.

[76] A. R. Newton and A. L. Sangiovanni-Vincentelli, "Relaxation-based electrical simulation," *IEEE Transactions on Electron Devices*, vol. 30, no. 9, pp. 1184–1207, September 1983.

[77] Nvidia CUDA, `http://www.nvidia.com/object/cuda_home_new.html`.

[78] Nvidia GTX 580, `http://www.geforce.com/Hardware/GPUs/geforce-gtx-580/specifications`.

[79] A. Odabasioglu, M. Celik, and L. T. Pileggi, "Prima: Passive reduced-order interconnect macromodeling algorithm," *IEEE Transactions on Computer-Aided Design of Integrated Circuits and Systems*, vol. 17, no. 8, pp. 645–654, August 1998.

[80] P. Odent, L. Claesen, and H. De Man, "A combined waveform relaxation-waveform relaxation Newton algorithm for efficient parallel circuit simulation," in *Proceedings of the European Design Automation Conference (EDAC)*, pp. 244–248, March 1990.

[81] J. M. Ortega and W. C. Rheinboldt, *Iterative Solution of Nonlinear Equations in Several Variables*. Academic Press, 2000.

[82] D. Paul, N. M. Nakhla, R. Achar, and M. S. Nakhla, "Parallel simulation of massively coupled interconnect networks," *IEEE Transactions on Advanced Packaging*, vol. 33, no. 1, pp. 115–127, February 2010.

[83] D. Pederson, "A historical review of circuit simulation," *IEEE Transactions on Circuits and Systems*, vol. 31, no. 1, pp. 103–111, January 1984.

[84] H. Peng and C. Cheng, "Parallel transistor level circuit simulation using domain decomposition methods," in *Asia and South Pacific Design Automation Conference (ASP-DAC 2009)*, pp. 397–402, January 2009.

[85] D. C. Pham, T. Aipperspach, D. Boerstler, M. Bolliger, R. Chaudhry, D. Cox, P. Harvey, P. Harvey, H. Hofstee, C. Johns, J. Kahle, A. Kameyama, J. Keaty, Y. Masubuchi, M. Pham, J. Pille, S. Posluszny, M. Riley, D. Stasiak, M. Suzuoki, O. Takahashi, J. Warnock, S. Weitzel, D. Wendel, and K. Yazawa, "Overview of the architecture, circuit design and physical implementation of a first-generation cell processor," *IEEE Journal of Solid-State Circuits*, vol. 41, no. 1, pp. 179–196, January 2006.

[86] D. C. Pham, T. Aipperspach, D. Boerstler, M. Bolliger, R. Chaudhry, D. Cox, P. Harvey, P. M. Harvey, H. P. Hofstee, C. Johns, J. Kahle, A. Kameyama,

J. Keaty, Y. Masubuchi, M. Pham, J. Pille, S. Posluszny, M. Riley, D. L. Stasiak, M. Suzuoki, O. Takahashi, J. Warnock, S. Weitzel, D. Wendel, and K. Yazawa, "Overview of the architecture, circuit design, and physical implementation of a first-generation cell processor," *IEEE Journal of Solid-State Circuits*, vol. 41, no. 1, pp. 179–196, January 2006.

[87] T. L. Pillage, R. A. Rohrer, and C. Visweswariah, *Electronic Circuit and System Simulation Methods*. McGraw-Hill, 1995.

[88] R. E. Poore, "GPU-accelerated time-domain circuit simulation," in *IEEE Custom Integrated Circuits Conference (CICC '09)*, pp. 629–632, September 2009.

[89] N. Rabbat, A. Sangiovanni-Vincentelli, and H. Hsieh, "A multilevel Newton algorithm with macromodeling and latency for the analysis of large-scale nonlinear circuits in the time domain," *IEEE Transactions on Circuits and Systems*, vol. 26, no. 9, pp. 733–741, September 1979.

[90] M. W. Reichelt, A. Lumsdaine, and J. K. White, "Accerlerated waveform methods for parallel transient simulation of semiconductor devices," in *ICCAD*, pp. 270–274, November 1993.

[91] M. Rewieński, "A perspective on fast-SPICE simulation technology," in *Simulation and Verification of Electronic and Biological Systems*, (S. Li and Feldmann, eds.), Springer, 2011.

[92] M. Rewienski and J. K. Kerns, "Optimization for post-layout arrays of cells for acclelerated transistor level simulation," United States Patent Application Publication, April 2009.

[93] D. L. Rhodes and A. Gerasoulis, "Scalable parallelization of harmonic balance simulation," in *IPPS/SPDP Workshops*, pp. 1055–1064, 1999.

[94] D. L. Rhodes and A. Gerasoulis, "A scheduling approach to parallel harmonic balance simulation," *Concurrency: Practice and Experience*, vol. 12, no. 2–3, pp. 175–187, 2000.

[95] D. L. Rhodes and B. S. Perlman, "Parallel computation for microwave circuit simulation," *IEEE Transactions on Microwave Theory and Techniques*, vol. 45, no. 5, pp. 587–592, May 1997.

[96] M. Rusu, J. Stinson, S. Tam, J. Leung, H. Muljono, and B. Cherkauer, "A 1.5-GHz 130-nm Itanium 2 processor with 6-MB on-die L3 cache," *IEEE Journal of Solid-State Circuits*, vol. 38, no. 11, pp. 1887–1895, November 2003.

[97] K. A. Sakallah and S. W. Director, "SAMSON2: An event driven VLSI circuit simulator," *IEEE Transactions on Computer-Aided Design of Integrated Circuits and Systems*, vol. 4, no. 4, pp. 668–684, October 1985.

[98] R. A. Saleh, K. A. Gallivan, M.-C. Chang, I. N. Hajj, D. Smart, and T. N. Trick, "Parallel circuit simulation on supercomputers," *Proceedings of the IEEE*, vol. 77, no. 12, pp. 1915–1931, December 1989.

[99] B. N. Sheehan, "Realizable reduction of RC networks," *IEEE Transactions on Computer-Aided Design of Integrated Circuits and Systems*, vol. 26, no. 8, pp. 1393–1407, August 2007.

[100] J. Shi, Y. Cai, W. Hou, L. Ma, S. X.-D. Tan, P. Ho, and X. Wang, "GPU friendly fast Poisson solver for structured power grid network analysis," in *ACM/IEEE Design Automation Conference (DAC '09)*, pp. 178–183, July 2009.

[101] H. Shichman, "Integration system of a nonlinear network analysis program," *IEEE Transactions on Circuit Theory*, vol. CT-17, no. 3, pp. 378–386, August 1970.

[102] H. Shichman, "Integration system of a nonlinear transient network-analysis program," *IEEE Transactions on Circuit Theory*, vol. 17, no. 3, pp. 378–386, August 1970.

[103] B. Smith, P. Bjørstad, and W. Gropp, *Domain Decomposition: Parallel Multi-level Methods for Elliptic Partial Differential Equations*. Cambridge University Press, 1996.

[104] K. Sun, Q. Zhou, K. Mohanram, and D. C. Sorensen, "Parallel domain decomposition for simulation of large-scale power grids," in *IEEE/ACM International Conference on Computer-Aided Design (ICCAD 2007)*, pp. 54–59, November 2007.

[105] Synopsys HSPICE, `http://www.synopsys.com/Tools/Verification/AMSVerification/CircuitSimulation/HSPICE`.

[106] H. K. Thornquist, E. R. Keiter, R. J. Hoekstra, D. M. Day, and E. G. Boman, "A parallel preconditioning strategy for efficient transistor-level circuit simulation," in *IEEE/ACM International Conference on Computer-Aided Design–Digest of Technical Papers (ICCAD 2009)*, pp. 410–417, November 2009.

[107] S. Tomov, R. Nath, H. Ltaief, and J. Dongarra, "Dense linear algebra solvers for multicore with GPU accelerators," in *IEEE International Symposium on Parallel Distributed Processing, Workshops and Phd Forum (IPDPSW)*, pp. 1–8, April 2010.

[108] *Using hierarchy and isomorphism to accelerate circuit simulation,* White Paper, Cadence Design Systems, Inc., 2004, `http://w2.cadence.com/whitepapers/5084_AccelCircuitWP_FNL.pdf`.

[109] S. Vangal, J. Howard, G. Ruhl, S. Dighe, H. Wilson, J. Tschanz, D. Finan, P. Iyer, A. Singh, T. Jacob, S. Jain, S. Venkataraman, Y. Hoskote, and N. Borkar, "An 80-tile 1.28 TFLOPS network-on-chip in 65 nm cmos," in *IEEE International Solid-State Circuits Conference (ISSCC)*, pp. 98–99, February 2007.

[110] *Virtuoso Multi-mode Simulation Datasheet,* Cadence Design Systems, Inc., `http://www.cadence.com/rl/Resources/datasheets/virtuoso_mmsim.pdf#page=7`, 2011.

[111] M. Vlach, "LU decomposition and forward-backward substitution of recursive bordered block diagonal matrices," *IEEE Proceedings G, Electronic Circuits and Systems*, vol. 132, no. 1, pp. 24–31, February 1985.

[112] A. Vladimirescu, *LSI Circuit Simulation on Vector Computers*. ERL-M82/75, University of California, Berkeley, CA, USA, 1982.

[113] D. M. Webber and A. L. Sangiovanni-Vincentelli, "Circuit simulation on the connection machine," in *Conference on Design Automation*, pp. 108–113, June 1987.

[114] W. Weeks, A. Jimenez, G. Mahoney, D. Mehta, H. Qassemzadeh, and T. Scott, "Algorithms for ASTAP — a network-analysis program," *IEEE Transactions on Circuit Theory*, vol. 20, no. 6, pp. 628–634, November 1973.

[115] Y.-C. Wen, K. Gallivan, and R. Saleh, "Improving parallel circuit simulation using high-level waveforms," in *IEEE International Symposium on Circuits and Systems (ISCAS '95)*, vol. 1, pp. 728–731, April 1995.

[116] D. F. Wendel, R. Kalla, J. Warnock, R. Cargnoni, S. G. Chu, J. G. Clabes, D. Dreps, D. Hrusecky, J. Friedrich, S. Islam, J. Kahle, J. Leenstra, G. Mittal, J. Paredes, J. Pille, P. J. Restle, B. Sinharoy, G. Smith, W. J. Starke, S. Taylor, J. Van Norstrand, S. Weitzel, P. G. Williams, and V. Zyuban, "Power7^{TM}: A highly parallel, scalable multi-core high end server processor," *IEEE Journal of Solid-State Circuits*, vol. 46, no. 1, pp. 145–161, January 2011.

[117] U. Wever and Q. Zheng, "Parallel transient analysis for circuit simulation," in *Proceedings of the Hawaii International Conference on System Sciences*, vol. 1, pp. 442–447, January 1996.

[118] J. K. White and A. L. Sangiovanni-Vincentelli, *Relaxation Techniques for the Simulation of VLSI Circuits.* Boston Norwell, MA, USA: Kluwer Academic Publishers, 1987.

[119] X. Ye, W. Dong, and P. Li, "MAPS: A multi-algorithm approach to parallel circuit simulation," in *ACM International Workshop on Timing Issues in the Specification and Synthesis of Digital Systems*, February 2008.

[120] X. Ye, W. Dong, P. Li, and S. Nassif, "Maps: Multi-algorithm parallel circuit simulation," in *IEEE/ACM International Conference on Computer-Aided Design (ICCAD 2008)*, pp. 73–78, November 2008.

[121] X. Ye, W. Dong, P. Li, and S. Nassif, "Hierarchical multialgorithm parallel circuit simulation," *IEEE Transactions on Computer-Aided Design of Integrated Circuits and Systems*, vol. 30, no. 1, pp. 45–58, January 2011.

[122] X. Ye and P. Li, "On-the-fly runtime adaptation for efficient execution of parallel multi-algorithm circuit simulation," in *IEEE/ACM International Conference on Computer-Aided Design (ICCAD 2010)*, pp. 298–304, November 2010.

[123] X. Ye and P. Li, "Parallel program performance modeling for runtime optimization of multi-algorithm circuit simulation," in *ACM/IEEE Design Automation Conference (DAC 2010)*, pp. 561–566, June 2010.

[124] H. Yoshida, S. Kumagai, I. Shirakawa, and S. Kodama, "A parallel implementation of large-scale circuit simulation," in *IEEE International Symposium on Circuits and Systems*, vol. 1, pp. 321–324, June 1988.

[125] C.-P. Yuan, R. Lucas, P. Chan, and R. Dutton, "Parallel electronic circuit simulation on the ipsc system," in *Custom Integrated Circuits Conference, 1988., Proceedings of the IEEE 1988*, pp. 6.5/1–6.5/4, May 1988.

[126] Z. Zeng, Z. Feng, P. Li, and V. Sarin, "Locality-driven parallel static analysis for power delivery networks," *ACM Transactions on Design Automation Electronic Systems*, vol. 16, pp. 28:1–28:17, June 2011.

[127] Z. Zeng and P. Li, "Locality-driven parallel power grid optimization," *IEEE Transactions on Computer-Aided Design of Integrated Circuits and Systems*, vol. 28, no. 8, pp. 1190–1200, August 2009.

[128] Z. Zeng, P. Li, and Z. Feng, "Parallel partitioning based on-chip power distribution network analysis using locality acceleration," in *Quality of*

Electronic Design (ISQED 2009) Quality Electronic Design, pp. 776–781, March 2009.

[129] Z. Zeng, X. Ye, Z. Feng, and P. Li, "Tradeoff analysis and optimization of power delivery networks with on-chip voltage regulation," in *ACM/IEEE Design Automation Conference (DAC 2010)*, pp. 831–836, June 2010.

Lightning Source UK Ltd.
Milton Keynes UK
UKOW05f1208080714

234746UK00001B/106/P

9 781601 985149